Recent Trends in Spices and Medicinal Plants Research

Recent Trends in Spices and Medicinal Plants Research

Amit Krishna De

2024

Associated Publishing Company®

A Division of

Astral International Pvt. Ltd.

New Delhi – 110 002

First Published, 2000

Reprinted, 2024

ISBN: 978-81-85211-47-3 (HB)

Disclaimer:

Every possible effort has been made to ensure that the information contained in this book is accurate at the time of going to press, and the publisher and author cannot accept responsibility for any errors or omissions, however caused. No responsibility for loss or damage occasioned to any person acting, or refraining from action, as a result of the material in this publication can be accepted by the editor, the publisher or the author. The Publisher is not associated with any product or vendor mentioned in the book. The contents of this work are intended to further general scientific research, understanding and discussion only. Readers should consult with a specialist where appropriate.

Every effort has been made to trace the owners of copyright material used in this book, if any. The author and the publisher will be grateful for any omission brought to their notice for acknowledgment in the future editions of the book.

Published by : **Associated Publishing Company®**
A Division of
Astral International Pvt. Ltd.
– ISO 9001:2015 Certified Company –
4736/23, Ansari Road, Darya Ganj
New Delhi-110 002
Ph. 011-43549197, 23278134
E-mail: info@astralint.com
Website: www.astralint.com

Printed at : Replika Press Pvt. Ltd.

CONTENTS

FOREWORD

Spices and medicinal plants are very important in human health and disease. Spices, apart from serving a crucial role in Indian culinary and export economy, are now revealing important properties in preventing or combating ailments including cancer, arthritis, metabolic and immunological disorders. Medicinal plants have been used since ages in tribal and folk medicines and Unani and Sino-Tibetan systems. Even in modern day Western Allopathic system, plants provide about one third of the medicines or the patent compounds.

Recently, multinational drug companies of developed countries are paying more and more attention to herbal medicine for their low cost, more sustainable curative value and less or on side effects. The recent controversy over the patenting of healing property of turmeric in America (and subsequent revoking of patent after strong representation by CSIR) and a number of patents on Neem are examples of the importance and economic use of Indian herbs and spices for human health care. In the context of U. N. declaration of "Health for All" by the year 2000 A. D., medicinal plants are expected to play a greater role. It is therefore very significant that this topic has been chosen, keeping in view all aspects of spices and medicinal plants.

India with its vast tropical and Himalayan biodiversity, medicinal plants has recently attracted renewed world attention in search of novel bioactive substances. On the other hand, lack of public awareness coupled with a leaning towards western lifestyle and dependence more and more on synthetic medicines are leading to loss of both faith in such systems of medicine and also of bio-diversity. Lack of proper trained and experienced herbal doctors (*vishaka*) and strict standardisation and drug rules for herbal medicines complicates the stituation.

With the free economy, acceptance of gene patenting and intellectual property rights, it is now high time to assess, document and conserve such valuable gift of nature as heritage for the benefit of man. Different genetic stocks and species of medicinal plants are to be identified using modern techniques and all out effort should be made to observe them on a war footing, before it is too late. This was the objective of the symposium on **Recent Trends in Spices and Medicinal Plants Research** where scientists, and, trade related experts in herbs and spices participated in the three day deliberations, to adequately address these questions and recommend means to achieve the goal.

I am happy that some selected papers of the symposium are being published as a permanent record. I would like to thank the Academy of Plant Sciences, India for giving the privilege of organising the symposium and the annual meeting of the Academy and allowing us to publish some papers presented at the symposium. I take this opportunity to thank Prof P. K. Ray, Bose Institute, Calcutta for extending all possible help to hold this meeting in Bose Institute. I also thank Dr. A. K. De, Organising Secretary of the symposium for editing this book, Prof B. R. Roy, Chairman Scientific Committee, and the publisher for making this publication possible. The co-operation of the speakers in submitting their manuscripts is also thankfully acknowledged.

Professor Radha Kanta Mandal
Bose Institute, Calcutta

PREFACE

The hills and oceans of India contain a variety of medicinal plants, which have been used traditionally for treatment of various diseases. Most of these plants are known to the local tribes and remains scientifically unexplored. Medicinal uses of quite a number of plants have been mentioned in vedas, puranas and susruta. From time immemorial, human beings have leaned on plants/herbs for curative treatment of diseases. Some of these plants can be grouped together as spices and they are regularly consumed by Indians not only for their nutritive properties but because of their aroma and flavour. It is time, that a thorough study of this valuable heritage is undertaken.

Books on medicinal plants or spices are available in the market but a book dealing together with both medicinal plants and spices is quite new. This book contains a variety of topics and is the first of its kind that deals exclusively with traditional & medicinal uses, cultivation, preservation, processing and trade of not only medicinal plants but also spices. Some aspects related to micropropagation, tissue culture and rural economy have also been considered. Moreover, this book provides valuable information on the current status of research related to herbal medicines throughout the country and tries to explain the folklore and traditional uses in terms of modern scientific findings.

Most of the articles published in this book from eminent scientists have been presented at the National Symposium on *Recent Trends in Spices & Medicinal Plants Research,* organised by Academy of Plant Sciences, India (APSI) in collaboration with APEDA and Spices Board from 2 to 4 April, 1998 at Bose Institute, Calcutta. It is hoped that this publication will enable others, who could not be present in the meeting, to know more about recent developments in the area of medicinal plants and spices. It is also hoped that students, teachers researchers, practitioners of traditional herbal medicine and commercial houses manufacturing drugs from natural ingredients, will find this publication very useful. The data in this book will be specially of use to pharmacists, analysts, botanists, toxicologists, chemists, pharmacognosists and biochemists.

Special acknowledgement is being made to APSI, APEDA and Spices Board for giving permission to include the papers in this book. The editor is also indebted to Mr. C K George and Mr P S Thampi of Spices Board for their co-operation, Dr. S K Gupta, Secretary, APSI for his valuable suggestions and to Mr S K Dutta of Associated Publishing Co. for taking the initiative to publish this book. I am also grateful to Professor Radha Kanta Mandal, Chairman of the Organising Committee of the above Conference for his constant inspiration and also writing the Foreword.

Editor

Chapter 1

Traditional and Medicinal Uses

Recent Trends in Spices and Medicinal Plants Research pp. 3-18, 2000
Ed. Amit Krishna De
Associated Publishing Co., New Delhi

1.1

SPICES AND CONDIMENTS: A BRIEF REVIEW

Amit Krishna De*

Abstract

The history, chemistry, pharmacology and trade of spices have been discussed briefly in this review. The traditional medicinal properties of some important spices have been stressed with recent research findings.

Introduction

What exactly spices are? Encyclopaedia Americana defines spices as "Products of plant origin for seasoning food to give flavour and aroma". Chambers Dictionary defines as "aromatic and pungent vegetable substances used as condiments and for seasoning food". The American Spice Trade Association (ASTA) defines spice as "any dried plant product used primarily for seasoning purpose". The word 'Spice', is derived from the Latin word 'species' meaning 'kinds of goods' while condiment derived from Latin word 'condire' means 'to preserve'. However, strictly speaking, spices are not food as they have very little nutritive value. Their role is to offer a pleasant flavour and aroma to the food and add greatly to the pleasure and taste of eating. They stimulate the appetite and augment the flow of the salivary and gastric juices. They are thus the food accessories or adjuncts. The food of man is to be seasoned by the addition of flavouring agents to make it more palatable and varied. This work is done by spices and condiments. According to the International Organization for Standardisation, there is no clear cut division between spices and condiments and so they are clubbed together. The term spices or condiments applies to 'such natural plant or vegetable products of mixtures thereof, in whole or ground form, as are used for imparting flavour, aroma and piquancy to and for seasoning food'.

Classification of Spices

Classifying the spices and to propose strict demarcating lines among them is very difficult. All aromatic vegetables used for flavouring foods as per definition, can be

* Indian Science Congress Association, 14 Dr. Biresh Guha Street, Calcutta 700 017.

classified under spices. But this is too wide a decision and would include many vegetables that are not in the true sense a spice. Other people restrict the term spices to the hardened and preservable parts of the plant that are usually used in the powdered form. They define condiments as spices or other flavouring substances that have a sharp taste and are usually added to the food after it is cooked. However, one way to go about in order to group the hundred of substances that are used today as spices, condiments, seasoning and flavouring materials is to take into consideration their botanical nature viz. roots, fruit seeds, leaves, whole herbs and so on. Some of the roots or underground parts that are used as spices includes ginger, turmeric, onion etc. While some of the barks are cinnamon, cassia, fruits may include nutmeg, cardamom, pepper, chilli etc. Tejpat, curry leaves, parsley may be from leaves and cumin, poppy seeds, basil, marjoram, rosemary are actually the seeds. The flowers includes cloves, hyssop etc. Classification of spices are also made according to their botanical source (families.)

There are about 70 species of spices grown in different parts of the world. Many of them are grown in India and other Asian countries. Around 52 spices have been identified by the Indian Spices Board till now. Recently, spices have been classified in terms of their origin and active principles present in them. The modern scientific instruments have been able to identify most of the active components of major spices. In this classification, six major categories have been made.
They are :

(i) Pungent spices e.g. pepper, chilli, ginger, mustard, etc.
(ii) Aromatic fruits and seeds e.g. cardamom, nutmeg, mace, fenugreek, caraway.
(iii) Aromatic bark e.g. cinnamon, cassia.
(iv) Phenolic spices e.g. all spices, clove.
(v) Coloured spices e.g. saffron, turmeric.
(vi) Umbelliferous fruits and seeds—aniseed, caraway, celery, coriander, cumin, dill.

Herbs are more domestic flavouring components deriving from the leaves of soft stemmed plants grown in temperate regions of the world. They contain relatively low percentages of essential oil. They are used either fresh or dried, preferably the former, for full flavour effect. Culinary herbs are generally classified according to their principal flavouring components. For example, herbs containing eugenol, thymol, thujone, mint and the sweet herbs. In these groups can be fitted the well known herbs such as marjoram, sage and numerous others. These and other non-culinary herbs have also been given medicinal attributes at different times. Allied to spices and herbs, some alliaceous vegetables (belonging to the genus *Allium*) like onion, garlic are also included as spices.

History of Spices

From the hieroglyphics, on the walls of the Pyramids to the scriptures of the Bible, spices have been mentioned repeatedly, particularly their role played in the life of the ancients. Pliny and Ptolemy have mentioned about spices and spicy land in their writings. The ancient commercial tie in spices existed with the Phoenicians, the

Chinese, the Persians, the Egyptians, the Romans and the Greeks. Mention of pepper was found in the old Testament which was perhaps more than 5000 BC old. Manna was described as being white as coriander seed. Clove was described in the legendary stories of 'Arabian Nights'. The famous stories about the voyages of Sindbad, the sailer, had numerous writings about the Malabar and its pepper. In his fourth voyage, it has been mentioned that he found men carrying pepper from the east.

Some important events that have occurred in the history of spices before the birth of Jesus Christ includes those like (i) Feeding of onions and garlic to one hundred labourers who toiled in the construction of the great Pyramid of Cheops, as medicinal herbs to preserve their health (2590-2568 BC) (ii) Camel caravans brought spices from Arabia Felix (1000 BC) (iii) The Queen of Sheba brought plenty of spices on her visit to King Solomon (950 BC) (iv) Ashurbanipal, king of Assyria established an ancient library at Nineveh in which many spices and herbs were described (700 BC) (v) Essential oils were produced from coriander and saffron and spices became a popular condiment in Persia (550 BC) (vi) Most of the writings of Charaka and Susruta contained uses of spices in medicine (500 BC) (vii) Hippocrates, known as father of Medicine, wrote many treatises in medicinal plants that included, saffron cinnamon, thyme, coriander, mint and marjoram (460-377 BC) (viii) The Port of Alexandria, founded by Alexander the Great, became central market place for spices, (322 BC) (ix) The Romans built ships and broke Arab monopoly of Indian trade (100 BC).

From ancient time, much of the trade in spices, particularly pepper, had been in the hands of the Arabs. In 637 A. D, an Arab fleet visited the West Coast of India and the Arab merchants were permitted by Hindu rulers to continue spice trade in Malabar. During the middle age, in Kerala, several kingdoms were prominent. Among them Cochin, Kozhikode, Cannanore, Cranganore, Quilon, Travancore, etc. were ruled by separate kings. These kings were deeply interested in spice trade as they got revenue by levying custom duties ranging from 2.5 percent to 6 percent on exports and transports of all spices both by land and sea.

Cochin port was considered as 'Venice of the East' in view of its natural beauty and commanding position on the ocean trade. The principal traders at Cochin were Muslims from Arabia and Egypt. They went by the name of 'Moors'. The Konkanis, a Hindu community from the Konkan districts, worked as spice trade agents. The king of Calicut was also the most powerful among all the coasts and for this distinction was called Zamorin (as much as Emperor). Cannanore was also a great emporium of spice trade and the king carried on brisk trade in spices with Moors and Arabs. The Arabs and the Persians had spice trade connections at Cannanore in the 12th and 13th centuries. During the period of the Tan dynasty, the Chinese had prosperous trade settlements at Quilon and during the reign of Kublai Khan in 13th century these two kingdoms exchanged envoys.

Marco Polo, a native of Italy came to India in 1292 and wrote an account of his travels. The fame of the Malabar coast as producer of spices was established in the West through the accounts of this Venetian traveller. The Moorish traveller Ibn Batuta in 1324, the Pope's Legate J. De Maanignoller in 1348, the Chinese Muslim traveller Mahuan in 1403, the Italian historian Nicolo Conti in 1410, the Mongolian ambassador Abdur Razzak of Heart in 1443 and the Russian traveller Athansius Niketan in 1468 visited Malabar and had nothing but praise for the spice trade. By the middle of the year 1429, during the reign of Henry-II, pepper merchants were established in London.

The Arabian merchants brought Malabar spices from Persian gulf ports after these had been ferried across the Arabian sea from West Coast of India. These spices were exchanged from Turkey to the European merchants. The bulk of the Indian spices were shipped through Constantinople for many centuries (from AD 395). But when Constantinople fell in the hands of Turks and this route was closed in AD 1453, a different route through Alexandria was sorted out. However, this route became unsuitable and Europeans started exploring different sea routes to reach India.

The king of Portugal sent Bartholomew Diaz in 1486 who reached southern tip of Africa called the Cape of Good Hope in 1487. Christopher Columbus, a Genoese sailer set out in 1492 to discover the Bahamas in West Indies which he mistook for India. Vasco da Gama studied the routes of other sailors and undertook sea voyage to Malabar. He came to Zanzibar along the east coast of South Africa and in 1498 reached near Calicut. Here he loaded his ships with spices and returned to Lisbon. He was greeted cordially by the king who conferred on him the title 'Dom'.

After this discovery of new sea route, Portuguese started trading spices from India. Centres were made at Cochin, Cannanore and Quilon. In 1516, a treaty was made with king of Quilon for spice trade which was renewed in 1544. However, the monopoly of spice trade which the Portuguese enjoyed in the Malabar coast encouraged the other European nations to start spice trade in India by the end of the 16th century.

In 1583, the English came from England to Malabar for spice trade. The East India Company was formed in 1600 and got a Charter from Queen Elizabeth for the purpose of carrying out trade with India. During 1616 and 1618, England imported on an average more than 14,32,000 lbs of pepper per annum.

In 1583, the Dutch traveller Van Linchoten came from Amsterdam to India. He wrote elaborately about the spice trade in India and this induced Dutch merchants to start spice trade. The Dutch East India Company was formed in 1594 and the United East India Company of Netherlands was formed in 1602. The Dutch tried their best to expel Portuguese from the Malabar coast and set up monopoly of spice trade in India.

The Portuguese slowly withdrew themselves from this competition and the Dutch and English established themselves. Coen was appointed the Dutch Governor General in the year 1618. In 1642, the Dutch made trade settlements with the king of Purakked and treaty with the king of Porka for pepper trade. They also expelled the Portuguese from Malabar and started enjoying the whole monopoly of trade in spices. The first treaty between the king of Cochin and Dutch East India Company was made in 1663 for trade of pepper and cinnamon. Amsterdam became the centre of spice trade in Europe. In 1866, a treaty was signed between the king of Travancore and Captain Nieuhoff, that nobody would be able to sell or export spices from Travancore State except the Dutch Company. Thus, in the 17th century, the Dutch enjoyed the supremacy in spice trade. From the middle of the 18th century, the British took over from the Dutch and in 1796 ousted them from Cochin. The English started trade in pepper from Tellicherry also. Slowly, the English established their political supremacy and the spice trade also came under their active control (Anonymous, 1993 a, b).

Cultivation of Spices

Spices have been cultivated and used since ancient times—from the beginning of human civilisation in India, China, Babylon, Egypt, Greece and Rome. Most of the spices originated in the tropical regions of Asia. The Arabs used to receive products from South India and was later superseded by the Dutch who retarded the cultivation of spices in principal spice producing countries. So severe, was the Dutch law with regard to cinnamon in Ceylon that the offence of cultivating a cinnamon tree on private land or destroying one belonging to the government was punishable by death. Cloves and nutmegs in the Moluccas were similarly guarded by Dutch law, the plants being either destroyed or their cultivation enforced to suit the circumstances. However, smuggling and migratory birds gradually introduced spices cultivation in other parts of the world.

Most of the spices of the world are produced in tropical Asian countries. Asia produces pepper, cardamom, cinnamon, nutmeg, mace, cloves, turmeric, ginger, and chilli. Other spices come from different countries e.g. the grains of paradise (viz. the aromatic seeds of *Aframomum melegueta*) is from Africa, while the vanilla (from the unripe but fully grown fruits of a climbing orchid *Vanilla planifolia*) from South America, red pepper or capsicum and all spice (dried unripe fruits of *Pimenta dioica*) from West Indies and Central and South America. A number of spices also occur in cooler temperate regions of the world. From the cooler regions of northern Europe and Asia, small quantities of coriander, cumin, caraway seed, mustard and calamus roots are obtained. Spice plants are mostly cultivated in tropical East Africa. It is, however, difficult to trace the origin of the spice yielding plants, though attempts have been made from time to time to find out their native homes. Cultivation of spices can, roughly be, classified into the following groups:

(i) Plantation cultivation—generally done on a large scale producing permanent crops for a number of years e.g, nutmegs, cloves, cardamom, cinnamons.

(ii) Garden cultivation—effected on a small scale and often grown as a subsidiary crop e.g. ginger, capsicum pepper.

(iii) Field crop—often grown in rotation with other crops e.g. Cumin and coriander.

The suitability of the type of soil for the proper growth varies greatly from spices to spices. In order to increase production, manures and fertilisers have to be applied, not only during the preparation of the field but also to be replenished from time to time to enrich the soil. The manures and fertilisers used may be grouped as follows (a) Animal manures—cattle dung and poultry dung, (b) Green manures—nodular leguminous plants, (c) Chemical fertilisers—inorganic materials like potassium, phosphorous and nitrogenous salts, and (d) special manures other than the above three types. However, the effect of application of fertilisers in India particularly for cultivation of spice plants is still under experimental stage and correct beneficial results can be estimated in near future.

Spices grown in Asia can be classified in terms of their production and export as: (a) Spices grown and exported in large quantities—black pepper, cardamom, turmeric, ginger, chilli, coriander, cumin, fennel, fenugreek, tamarind, garlic etc. (b) Spices grown in smaller quantities—anise, bayleaf, cinnamon, dill seed, poppy seed etc. (c) Spices grown in insufficient quantities—clove, nutmeg, mace, all spice, and herbal spices like rosemary, marjoram, sage, parsley etc. Around 8 million farmers are engaged in growing on 18 lakh ha to produce more than 2 million tons of spices annually.

Spices grown in West Bengal are chilli, ginger, turmeric, coriander, black cumin, large cardamom and a small quantity of black pepper. The chilli is cultivated in Sundarban area, West Dinajpur, Murshidabad, while the ginger and large cardamoms are grown mostly in Darjeeling and Jalpaiguri areas. Turmeric is cultivated mostly in Murshidabad, Jalpaiguri and Midnapore areas. There are about 20 varieties of chillies grown in the State of which, most popular are the Suryamukhi-1, Suryamukhi-2. They are mostly used as green chillies. Apart from these two varieties are Patnai, Suli, Siti, Sundari, Bullet etc. Black pepper is not very much grown in West Bengal, but is largely cultivated in Kerala and Karnataka along with small cardamom.

Black pepper is known as the 'Black gold of India'. Asia cultivated black pepper in tons and supplied 70-80 percent of the total world requirement, which has, however, come down to around 20 percent recently. The major pepper producing countries like Brazil, Indonesia and Malaysia have recently overtook India. Around 24 types of pepper are grown in South India, out of which Kerala contributes 95 per cent and Karnataka 3.5 per cent of India's total production.

The second most important spice of Asia is cardamom which is known as 'Queen of Spices' and comes next to pepper. India is the largest cardamom producing country in the world, the other important countries are Guatemala, Sri Lanka, Tanzania and Taiwan. Recently, Guatemala has overtook India in the export market of the world. Kerala accounts for 60 per cent of the world trade.

Among the other spices, Indian ginger is considered second to the Jamaican variety in quality. The major producer of ginger other than India are Taiwan, Jamaica, Nigeria and Malaysia. India supplies around 50 per cent of the world market. Turmeric is also cultivated in huge quantities in different parts of India, particularly in Kerala, Meghalaya, Andhra Pradesh, Tamil Nadu. Maharashtra and also in some parts of Bihar and Orissa. There are about 70 species of turmeric of which 30 occur in India.

Export and Trade

All parts of the world contribute spices. Thus, Asiatic tropics is the home of the greater part of the valuable spices—pepper, turmeric, ginger, celery, chillies, cassia, cinnamon, nutmeg, cloves, cumin, cardamom, etc. Pepper comes from Brazil, cardamom from Guatemala, mace and nutmeg from Grenada, cloves from Madagascar and cumin from Iran, Turkey and China. Vanilla hails from Bourbon Islands (Madagaskar, Comora, Reunion and Seychelles), Indonesia, Mexico and Tonga, paprika from Spain, Hungary, Morocco, Yugoslavia, USA. and Mexico. Spain, Hungary, Turkey and India provide saffron; Southern Europe and India provide fennel. Holland, Poland, Iran, Turkey and India give poppy seed; mustard is produced in Canada and Sesame in China, India, Africa, South America and the USA. The colder regions of the world supply coriander, caraway, rosemary and aniseed. On the whole, India is the largest producer—roughly 2.5 million tons and exports—roughly 1.76 million tons.

International trade in spices is estimated to be around 400,000 tons annually at approximately US $1.5 billion. The total value depends to a large extent on the prevailing price of pepper, which is the leading spice in the world market. This trade had expanded steadily over the past two decades. Spice exports were only slightly more than the present amount at 2,22,000 tons ($300 million) annually. From 1981 to 1985, the yearly average moved up to 3,50,000 tons ($1 billion). The major market for spices are the United States, Germany, Japan and some Middle East countries. The United States is the world's largest importer of spices with foreign purchase of 2,42,719 tons, valued at $395 million. The European community imports 2,02,483 tons of spices valued at $423 million. Germany is the foremost market in Europe and has the second world market after the United States. It accounts for one third of total spice imported into Western Europe estimating to 63,731 tons valued at $137 million. Next in line are France and the United Kingdom which imports 31,600 tons and 27,000 tons coming to $68.3 million and $44.4 million respectively. Among other European markets, the leading importers are Austria, Sweden, Norway, Finland and Switzerland.

In recent years, markets in the Middle East, in particular Saudi Arabia, have accounted for a substantial and increasing share of the spice trade, in terms of value, largely because of considerable imports of cardamom and pepper. These markets hold a share of over 80 percent of total world consumption of cardamom. Morocco, Algeria and Libya are large purchasers of pepper, but demand in these three countries is closely tied to price levels.

In Asia and the Pacific region, the major spice importer is Japan, the third largest market for spices in the world, followed by Australia and New Zealand. India is the largest producer and overall consumer of spices. Although domestic consumption in Singapore and Hong Kong is small, the importance of former, in particular, in the entrepot trade of spices still remains considerable. Pepper, mainly from Indonesia and more recently from Vietnam, passes through Singapore to the main consuming markets. It is estimated that around one fifth of the pepper entering international trade, moves through Singapore. Indonesia is a major producer and exporter of a number of spices such as pepper, cassia, nutmeg, ginger and largest user of cloves for the 'Kretek' cigarette industry. Indonesia, now produces enough clove to sell in the international market.

India, today, not only exports raw spices, but also various value added items to around 70 countries. As spices in the raw form have certain disadvantages for export they can be traded in variety of forms, like ground spices, and spice extracts that includes spice oils, oleoresins and derivatives of oils/oleoresins. There is a high demand of oils and oleoresins from Indian spices throughout the world. The Indian oleoresin industry is contributing for the major foreign exchange earning in India. During the last few years, the export of value added spices from India have shown significant growth and has registered a growth rate of 22 percent per annum.

Composition and Uses of Spices

Spices are obtained from a large number of different plants. They are parts of plant, such as roots, buds, flowers, fruits, barks or seeds. Thus little can be said, in a general way, about their composition. Most spices owe their flavouring properties to volatile oils and in some cases to fixed oils and small amount of resins which are known as oleoresins. In many cases, no single compound is responsible for flavours, a blend of different compounds such as alcohols, phenols, esters, terpenes, organic acids, resins, alkaloids and sulphur containing compounds contribute to the flavour. In addition to flavour contributing compounds, all spices contain the usual components of plant products, such as proteins, carbohydrates, fibre, minerals and tannins or polyphenols.

Spices being agricultural commodities, are prone to spoilage by insect or microbial attack. Hence, the spice oils or oleoresins which contain all the active ingredients of spices are extracted or marketed. Spice oils are obtained by steam distillation of ground spices. Oleoresins are obtained by solvent extraction of ground spices. The spice oils contain only the aromatic contents while the oleoresins contain both the aromatic and pungent properties. Various solvents like acetone, isopropanol, methanol, hexane, etc. are used as solvents. According to ISI standards, oleoresin obtained from solvent extraction should not contain more than a specified amount of solvent. The processed products of spices have several advantages—they are convenient to use, free from contamination, have better storage life and are easy to transport. Technology has developed in Asia to prepare ready to use oleoresin from spices.

Spices are mostly used as flavouring agents in a number of food products; such as curries, backery products, pickles, processed meats, beverages, liquors etc. They enhance or vary the flavour of foods. Spices are also flavour disguisers, they help mask off the flavour of foods, which if unspiced may have to be thrown away. Some spices possess antioxidant properties while others are used as carminatives. Some spices are mainly used as preservatives in some foods like pickles and chutney. Others, like clove and mustard possess strong anti-microbial properties and as such prevent food spoilage. Spices have been used since ages to preserve food specially meat and fish for long periods when there was no refrigeration. Many spices also possess important physiological and medicinal properties.

In allopathic medicine, the use of spices is not considered to be as important as they were once regarded in the Middle ages. Nonetheless, they are much prevalent in Ayurvedic system of medicine and many of them happen to be officially recognised drug materials in the pharmacopoeia of several countries in America as well as Europe, besides many other countries in the orient. They are used as carminatives (to expel gases from the digestive system), antiseptics (as they kill bacteria) and to mask the unpleasant taste of other drugs. In addition, they do constitute an important ingredient in many industries such as perfumery, soap manufacturing and also dyes. At one time, spices had been actually used in lieu of hard cash as well. The composition and some of the specific uses of major spices are given here :

(i) Mace and Nutmeg (*Myristica fragrans* Houtt.)

It contains 4-15% volatile oil which is responsible for its typical smell and flavour. The oil of mace contains 93% terpenes and 77% aromatics, while that of nutmeg contains 84% terpenes and 11% aromatics. The volatile oil comprises of camphene, di–pentene, eugenol, myristicin, safrole, iso–eugenol, geraniol, d-linalool, 1-terpinen-4-ol, β-pinene. Nutmeg also contains 29-30% fixed oil known as nutmeg butter that is orange in colour. The aromatic oil comprises of trimyristrin, fats, resin and traces of acid like formic, acetic, cerotic acid etc.

Recently, vanillic acid and synapic acid from nutmeg and mace respectively has been isolated. Nutmeg and mace has a variety of medicinal uses besides their culinary uses. It acts as stimulant and carminative. The oil is used in medicated soaps, dental pastes, flavouring chewing gums, chewing tobacco etc. It is used to great extent in perfumery, dentrifices and in tobacco industries. It has anti-inflammatory properties and is used specially for treatment of bladder and urinary tract inflammation. It is sometimes used to relieve excessive flatulence, dyspepsia, hysterical cough, fainting and sleeplessness. In homeopathy, it is used in treating hysterical women. In china nutmeg is used as carminative, stomachic and antispasmodic drug. In Cambodia nutmeg is recommended for treatment of looseness of the bowels.

(ii) Turmeric (*Curcuma longa* Linn.)

Turmeric contains essential oil of around 5.8% containing di-phellandrene, d-sabinene, cineol, borneol, and zingiberene. It also contains compesterol, stigmasterol, β-sittosterol, cholesterol, monocyclic sesquiterpene, ketone tumerone and p-totyl methyl carbinol and curcumin. Turmeric owes its colour primarily to curcuminoids ranging from 0.2 to 2% depending on the species and on pre and post harvest factors. The curcuminoids are curcumin (difeneloxylmethane; 1,7 bis(4 hydroxy 3-m methoxyphenol)-1,6-heptadiene 3,5-dione). Fatty acids are composed of saturated straight chain, saturated iso, monoenoic and dienoic acids demethoxy curcumin and bis demethoxy curcumin.

Turmeric is an important condiment used in cooking, specially in the form of curry powder. It is also used in cosmetics and as natural dye for colouring wool, silk, cotton, confectioneries and in food industries. It is antiseptic and has antimicrobial properties (Khanna, 1999). Curcumin is a potent antioxidant and can inhibit cancer. Studies in smokers have revealed that turmeric given in daily dose from a fortnight to a month can decrease the mutagens in urine. It has also been demonstrated to show anti tumor effects in animals treated with potent carcinogens. The pharmacological properties of curcumin includes anti-inflammatory, antioxidant, inhibition of platelet aggregation, local spasmolytic, hepatoprotective, and choleretic properties (6-8) (Piyachatu Rawat, 1996; Bonte, 1997). It also causes lowering of blood cholesterol levels.

(iii) Cardamom (*Elettaria cardamomun Maton/Amomum subulatum* Roxb.)

The principal constituents of volatile oil are cineol, α-terpineol, myrcene, terpinene, d-limonene, sabinene and terpineol acetate, sesquiterpenes, α & β-pinene, monoterpene hydrocarbons, 1,8-cineole, linalool acetate. The aqueous portion of the steam distillate contains 0.5% essential oil and one of the chief constituents is borneol. The component fatty acids of the fixed oil is caprylic, caproic, palmitic oleic and linoleic acids. β-sitosterol has also been reported to be obtained from the unsaponifiable fraction of the fatty oil.

Cardamom is mainly used as flavouring agent for sweet meats, beverages, and liquors. The oil from the seeds is used for treatment of inflammation of the eyelids. Cardamom is a carminative and is used together with betel leaf. The root and fruit of cardamom are used for its laxative and tonic properties. They are used as diuretic stimulant, stomachic and correcting menstrual disorders.

(iv) Clove (*Szzygium aromaticum* L. Merr. et. Perry; *Syn. Eugenia aromaticum*)

Clove contains a heavy volatile oil of 16-20%, a camphor resin 6%, caryophylin and eugenin—the alkaloids, woody fibre, gum etc. The main constituents of volatile oil are eugenol, eugenol acetate, caryophyllene, methyl n-amyl-carbinol, methyl furfural alcohol, dimethyl furfural, methyl-n-heptyl carbinol, α–methyl furfural, naphthalene, and vanillin. Clove also contains calcium, phosphorous, iron, sodium, potassium, thiamine, riboflavin, niacin and vitamin A and C. The clove buds, stem, and leaves on steam distillation yields substantial amounts of volatile oil. The bud contains free eugenol, eugenol acetate and caryophyllene. The stem oil contains more free eugenol than bud oil while the leaf oil contains least eugenol and eugenol acetate.

Clove is used either as whole or in powder form in preparing curry powder sauces, pickles, preservatives and ketch up. It is used for flavouring cakes, puddings, cookies, chocolate and candies. Clove oil is used in confectioneries. Besides being used in relieving toothache, headache and rheumatic pain, it is used as counter irritant. It has antiseptic, anti-inflammatory and analgesic properties. The oil is used as carminative, spasmodic and aids digestion. It is now used in soaps, detergents and in perfumary.

(v) Fennel (*Foeniculum vulgare* Milli)

The mineral contents of fehnel are calcium, phosphorous, iron, sodium, potassium and vitamin contents are thiamine, riboflavin, niacin and vitamin C. Its calorific value is 370. The composition of oil varies widely according to variety and origin. Indian fennel contains palmitic acid and petroselinic acid, d-α-pinene, camphene, d-α-phellendrene, limonene, anethole, anisaldehyde, p-anisic acid and dipentene d-fenchone. Methyl chavacol and foeniculin are also present as seed oils. Water extracts of the seeds shows the presence of thermostable growth retarding substance. The diuretic properties of fennel is attributed to the presence of glycollic acid.

Fennel is mostly used as flavouring agent in food. Young leaves of the tree are used raw or cooked with curry. The seeds are generally used for flavouring and the fruit is commonly used as a masticatory. The seed and the oil has a variety of medicinal properties. The leaves of fennel are digestive, appetizing and stimulating. They increase the secretion and discharge urine. The seeds are sweet, laxative, aphrodiasic and arrest bleeding. They also relieve flatulence and promote the removal of catarrhal matter and phlegm from the bronchial tubes. Oil of fennel, distilled from the dry seeds is aromatic, carminative and antispasmodic. It is used in various carminative preparations.

(vi) Ginger (*Zingiber officinale* Rosc)

It contains 1-3% essential oil with a characteristic pleasant and aromatic odour. The major constituents are monoterpenes, sesquiterpenes and oxygenated terpenes. The other constituents are—chovicol, zingiberene, zingiberol, n-decylaldehyde, n-nonylaldehyde, d-camphene, d-β-phellandrene, methyl heptanone, cineole, d-borneol, geraneol, linalool, citral, nonanol, heptane, octane, gingediol, methylgingediol and limonene. The non pungent principles are palmitic and other fatty acids while the pungent principles are oxymethyl phenols such as gingerol, shagoal, zingerone and related compounds.

Ginger is used in dried, preserved and green forms. Because of its pleasant aroma and pungency, it is widely used as flavouring agent in the preparation of food products like ginger bread, confectioneries, curry powder, curried meats, pickles, and soft drinks like ginger cocktail, gingerella and carbonated drinks like ginger beer. It is also used for the preparation of oleoresins and essential oils. Ginger oil finds wide use in perfumery. It has a variety of medicinal properties and are used as anti inflammatory and digestive agents. Ginger oleoresin when administered orally significantly lowered serum and hepatic cholesterol and increased faecal cholesterol excretion. It possesses anti-oxidant properties and may be added to edible oils and fats to protect them against rancidity. It is carminative, stimulant and given for flatulence and colic. It has also hypoglycemic activity (Singh and Singh, 1996).

(vii) Tejpat (*Cinnamomum tamala* Nees)

It is a flavouring agent. The flavour is because of an alcohol soluble essential oil containing 70-80% cinnamic aldehyde. Moreover, high amounts of d-phellandrene and eugenol are found among other constituents.

Besides flavouring agent, tejpat oil is used for perfuming soap and in medicines. It is used for treatment of boils, conjunctivitis, itching, diarrhoea, fatigue, tiredness and urinary problems.

(viii) Saffron (*Crocus sativus* Linn.)

It contains carotenoid glycosides β-carotene, α-carotene, zeaxanthin, lycopene, riboflavin and crocin. The glycosides crocin is the colouring material together with β-carotene. The essential oil of saffron contains stearoptene (a tertiary alcohol).

Mostly used for colouring and flavouring confectioneries. It is also used in dye industry. In homeopathic medicine, it is used for treatment of hysteria and sudden changes in moods. It is also known to provide strength and improve sexual life of males if taken with milk. Recently, an anti-tumor compound from saffron has been isolated that has antioxidant effects (Nair *et. al*, 1994).

(ix)Coriander (*Coriandrum sativum* Linn.)

It yields a colourless pale yellow oil, chief constituent of which is cinnamaldehyde, eugenol, benzaldehyde, methyl n-amyl ketone, rutin, linalool, d-z pinene, dl z pinene, p-cymene, dipentene, y-terpinene, phellandrene, terpinolene and traces of coriandrinol, angelicin,l-borneol, n-decyolaldehyde thymol, elemol, geraniol caryo phyllene, safrole, cinnamyl alcohol, methyleugenol. The constituent fatty acids of the oil are palmitic, petrosebinic and linolic acids. Besides, variety of minerals and vitamin, coriander contains high amount of vitamin A.

Coriander fruits are used as culinary preparations, flavouring liquors, cocoa preparations and in confectionery. It has medicinal properties and used for treatment of bronchial asthma, bronchitis, small pox and conjunctivitis. Coriander leaves are stimulant and tonic. Coriander bark oil has high germicidal activity and is used as fungicide.

(x) Asafoetida (*Ferula asafoetida* Lnn.)

The major constituents of asafoetida are moisture, ash, resin, gum glucuronic acid, galactose, arabinose, rhamnose. Volatile oil is 10-17% essential oil 6-10% containing sulfur compounds, umbelliferone, ferulic acid and pinene. Recently, phellopterin, isopimpinellin and byakangelicin has been isolated.

It is used as a medicine and as a condiment. The gum resin possesses antispasmodic, carminative, expectorant and stimulant properties. It is effective in curing respiratory problems, and cold attacks. It is used for treating hysteria, syphilis, excessive flatulence and bronchitis. It has antimicrobial properties. Essential oil shows protective action against fat induced increase in plasma fibrinogen and decrease in coagulation time and fibrinolytic activity on alimentary hyperlipaemia (Bakhu, 1992).

(xi) Fenugreek (*Trigonella foenum-graecum* Linn.)

It contains fatty acids, an alkaloid trignelline and a liquid base choline. The saponins of trignella contains sapogenin, diosgenin, gitogenin, tigogenin, reotigogenin, yanogenin. The non saponificable fraction contains carotene, ergosterol and phospholipids while saponifiable fraction contains a nonphenolic lactating factor. The alcoholic extracts of the plant also contains a high amount of an amino acid cystein. Glycosides, graecunins, fenugrin B, etc are also present in the seeds.

Fenugreek is used to increase the flow of milk of cattles as it contains a lactation promoting factor. Used to cure fatigue, small pox, gastric trouble, gout, flatulence, loss of appetite, irregular period in woman and sexual weakness. It is also used in colic flatulence, dysentery, diarrhoea, chronic cough, enlargement of spleen and liver, rickets and diabetes. Fenugreek reduces blood sugar levels in diabetic patients (Reghunathan *et. al*, 1994). It contains 'diosgenin' which is used as starting material in synthesis of sex hormones and oral contraceptives. It is often grounded and mixed with cotton seeds and fed to the cattles for increase in milk production. In Java, it is used in hair tonic for cure of baldness. In the dye industry the watery extract of the seed is used to import glass and durability.

(xii) Garlic (*Allium sativum* Linn.)

Allyl prophudisulphide, responsible for the smell, is the main constituent of essential oil of garlic. Other constituents of the oil are diallylsulphide and two more sulphur containing compounds. Allicin, allisatin-I, allisatin-II are the chief plant antibiotics obtained from it. Allicin is obtained from alline by the action of the enzyme allinase. Allylthiamin has been isolated from bulbs of garlic. The carbohydrates in garlic bulb consist mainly of invertase and polyfructosidase.

Garlic is used for flavouring curries, chutnies, soaps and stews. It also contains a variety of medicinal properties. It acts as stimulant, carminative, emmenagogue, antirheumatic, ant-helminthic and stimulant. It is also used as a verifuge i.e to expel worms, specially round worms. Garlic is known to lower cholesterol in the blood and is recommended for heart diseases and arthereosclerosis (Jain, 1972). Garlic also shows antibacterial and anti-fungal effects against a variety of strains (Alam *et. al.* 1996). The sedative effect of garlic has been attributed to a delaying action upon excessive motor activity of stomach and intestines. It has antiinflamatory activity and is effective in the treatment of rheumatoid arthritis (Singh *et. al.* 198). It is also hepato-protective (Singh *et. al.*, 1997).

(xiii) Mustard (*Brassica juncea* Linn. Boissier)

It contains the alkaloid sinapine and also myrosin, sinigrin and some colouring matters. Myrosin is a glucoside while sinigrin is a potassium myronte which when acted upon by water yields sulphocyamide of allyl which is the volatile oil of mustard that causes its unique odour. The seed also contains 20 per cent sinapine, sulphacyanide, lecithin, mucilage, B-phenylethyl isothiocyanates. The fixed oil obtained by expression contains glycerides of oleic, stearic and eracic or brassic acids.

Externally its oil is stimulant and a mild counter-irritant, Internally the seeds cause vomiting. In moderate doses they are digestive and laxative. Mustard plasters are beneficial in gout, sciatica and urticaria. It is useful in inflammatory swellings. e.g. parotitis. Expressed oil is used in cooking and is also applied externally in mild attack of sore throat, internal congestion and chronic muscular rheumatism.

(xiii) Dill (*Anethum graveolens* Linn.)

Dill oil contains (+) carvone as the major constituent of around 20-30% by volume. Besides this ketone, dillapiole (1 allyl 1-3, 4-methylendioxy-5, 6-dimethoxy benzene) tripetroselinin, petroselinicdiolein myriotrin, eugenol, pinene, limonene, thymol, phellandrene, terpinene are also present.

Dill seeds are carminative, stomachic, aromatic, stimulant, diuretic and controls menstrual flow. It is used frequently for insecticidal property due to presence of carvone. However, both dill seeds and dill leaves are mainly used for flavouring pickles, salads, tomato juice, curry powder etc.

(xiv) Pepper (*Piper longum* Linn.)

It contains a volatile oil, resin and an alkaloid piperine. The essential oil consists of α-pinene, β-pinene, 1-α-phellandrene, dl-limonene, piperonal, dihydrocarveol, α-caryophyllene and piperidine complex. The oleoresin contains piperine, chavacine, other piperidides pipercides and piperidine. The volatile alkaloid α-methyl pyrroline is also present.

Pepper is indispensible in sauces, soups, condiments, pickles and for seasoning dishes. It is stimulant, pungent, aromatic, digestive and appetiser. It increases flow of saliva and gastric juices and used for digestive disorders. Piperidine, piperine is a potent anti-inflammatory agent and pepper is used for gum inflammation, toothache, muscular pain. It is also potent analgesic. Pipercide shows insecticidal activity (Khajuria *et al*, 1997).

(xv) Chilli (*Capsicum frutescens/annuum* Linn.)

Chilli gives a fatty oil (9-13 per cent) and on steam distillation gives volatile oil (0.16-39 per cent). Chilli contains pungent principle capsaicinoids that is a mixture of capsaicin, dihydro capsaicin, homo-capsaicin, non-hydrocapsaicin, and homo-dihydrocapsaicin. Some pigments like capsanthin, capsorubin, zeaxanthin, lutein, cryptoxanthin and xanthophylls are also present. Chilli is rich in vitamin C. Seed oil contains citrostadienol, gramisterol, methyllophenol, ethyelophenol, lophenol, cycloeucalenol, obtusifoliol, etc.

Dry chilli is extensively used in curry powder and curried dishes. Capsaicin is a potent anti-inflammatory and analgesic agent. Chilli is used for treatment of headaches, toothaches, muscular sprains, low backache and sciatic pain, It has cholesterol lowering effects and is used as anti-obesity agent (De Ak, 1994). Recently, it has been found to be effective against gaseous irritant (even cigarette smoke) induced pulmonary damages (De and Jhash, 1988, 1992, 1993). It is also digestive and increases saliva flow. Chilli also has potent anti–microbial properties (Cichewicz and Thorpe, 1996).

(xvi) Tamarind (*Tamarindus indica* Linn.)

Fruits contain traces of oxalic acid while seeds contain a fatty oil. The pectin contains polysaccharide and xylan. The pulp contains tartaric and other acids.

Tamarind is used as an acidulant in some food preparations. It is digestive, antiflatulent, cooling, laxative and antiseptic. Tamarind pulp being rich in vitamin C is prescribed in scurvy, common cold and fevers. It is used in dyeing, tanning and polishing and cleaning metalware. Tamarind powder finds extensive use as a sizing material in textile industry.

(xvii) Onion (*Allium cepa* Linn.)

The chief constituent of the essential oil is allyl propyl disulphide. The volatile oil contains methanol, propionaldehyde, n-propyl-thiol hydrogen sulphide, acetaldehyde an-hydroxy propanthiol, 4-hexenyl aldehyde, cyclo-allin and α-glutamyl peptides-(1-proponyl)-cysteine sulphoxide is considered very important.

Fresh onions have a strong anti-microbial effect. Onions offer a considerable relief in neuralgia, headaches and migrane. It is also useful in dyspepsia or mild indigestion and excite appetite in food, removes common cold, overcome flatulence and improves digestion. Onions are found beneficial in inflammation, ulceration of mouth and gums and toothache. It is used in gout, constipation and urinary disorders.

(xviii) Bishops Weed (*Trachyspermum ammi* Linn. Sprague.)

The principle oil consists of thymol. It is stimulant, antispasmodic, tonic and carminative, administered in flatulence, dyspepsia, diarrhoea and often recommended in cholera. The oil is used for sore throat and bronchitis. It has antimicrobial activity and is used in soaps, plasticizers and in fertilisers (Mehta *et al*, 1994).

(xix) Cumin (*Cuminum cyminum* Linn.)

Seeds yield a volatile oil with an unpleasant bitter taste containing cuminaldehyde, p-cymene, pines, cuminyl alcohol, phellandrene, and l-terpineol. Moreover, glucosides of luteolin, apigenin and d-manitol are present in the seeds.

It is generally used as an additive in increasing flavour of different food preparations. It also has a variety of medicinal properties and used as carminative, astringent, stomachic and useful in dyspepsia and diarrhoea. It is used in perfumery and flavouring liquors.

(xx) Black Cumin (*Nigella sativa* Linn.)

It contains a volatile oil containing myristic acid and telfairic acids, sapogenin, melanthigenin.

The seeds are carminative, stimulant, appetiser, diuretic, emenagogue, galactogogue, anti–helminthic, and used for treatment of puerperal fever. The powdered seed mixture with sesame oil is often externally applied in the ruptions of the skin. Alcohol extracts showed anti-microbial activity. Recent experiments indicate that black cumin seeds have anti-carcinogenic activity. It is used as insecticide and can protect other seeds from insect attack when mixed. The oil is useful in cough and bronchial asthma.

(xxi) Basil (*Ocimum basilicum* Linn.)

It consists of essential oil mainly of methyl cinnamate, linalool and a terpenic constituent ocimene.

Basil is used for culinary purposes. It is very popular in french cooking. The juice of the leaves is given specially to infants for cough, cold and fever. It has antipyretic activity (Chatterjee TK, 1977). Basil is antiseptic and has antibacterial activity. The plant is considered stomachic, alexipharmac, antipyretic, disphoretic, expectorant, carmonative, stimulant and antihelminthic.

(xxii) Horse Radish (*Armoracia rustica* Gaerin)

It contains water 73.4%, protein 3.2%, fat 0.2%, ash 1.8%, total carbohydrates 21.4% and fibre 2.4%. The root contains a pungent acid vesicating volatile oil. The pungency is due to the glycoside sinigrin and the enzyme myrosin, which react when the root is crushed and mixed with water yielding allyl isothiocyanate. The root is also a rich source of ascorbic acid (vitamin C)

Horse radish is well known for its use as condiment and has been credited with digestive and antiscorbutic properties. It is also stimulant, diaphoretic, diuretic and digestive. It is used as counter irritant in lumbago and similar painful joints. Crushed horse radish has an inhibitory effect on the growth of microorganisms due to possible presence of allyl isothiocyanate.

(xxiii) Curry Leaf (*Murraya koenigii* Linn. Spren.gel)

Leaves and fruits contain alkaloids koenimbine, mahanimbine and koenigicine. Moreover, leaves contain mahanimbidine, cyclomahanimbin curryangine and curryanine. Marrayanine, mukoeic acid and girinimbine were also obtained from the bark. Curry leaves contain moisture 66.3%, protein 6.1%, fat 1.0%, carbohydrates 16.0%, fibre 6.4% and mineral matter 4.2% per 100 grams. The leaves are rich in calcium, phosphorous, iron, nicotinic acid and vitamin C. The leaves also contain a residual glucoside named koenigin.

Curry leaves possess the function of stomachic and is also a mild laxative. It is used in the treatment of morning sickness, nausea, flatulence and vomiting. It is a hair tonic and is considered beneficial in preventing premature greying of hair. The aqueous extract of leaves on oral administration has been shown to give hypoglycemic activity in normal and alloxan diabetic dogs (Chatterjee and Pakrashi, 1991).

(xxiv) Parsley (*Petroselinum crispum* Miller)

It contains moisture 74.6%, protein 5.9%, fat 1.0%, mineral 3.2%, fibre 1.8% and carbohydrates 13.5%. It contains a glucoside aplin and an essential oil which contains apiol. The fruit contains coumarin. The other constituents are apiole (1 allyl 2, 5-dimethoxy-3,4-methylene dioxybenzene), allyltetramethoxy benzene, myristicin and a mixture of flavone glycosides.

Parsley helps in digestion and genito–urinary problems. It is known to maintain blood pressure. It is good for weak eyes, ulceration of cornea, cataracts, conjunctivitis and opthalmia. Recently, it has shown antifertility activities in animals.

The other spices whose major chemical composition are still to be identified clearly and which have limited uses are rosemary, aniseed, cellary, kokam, mint, star anise, sweet flag, greater galanga, caper, cambodge, hissop, juniper berry, lovage, marhoram, all spice, sage, savory, thyme, and tarragon.

References

Alsam, M., Islam, S. N., 1996. Garlic: a broad spectrum antibacterial agent effective against common pathogenic bacteria. *Fitoterapia*, **67**: 374.

Anonymous, 1993 a. History of Indian Spices, *Spices India*, **5**(2), 2.

Anonymous, 1993 b. History of Indian Spices, *Spices India*, **6**(1), 2.

Bonte, F., Hudson, N., Wepierre, J, Meybeck, A., 1997 Protective effect of curcuminoids on epidermal skin cells under free oxygen radical stress, *Plant Medica*, **63**, 265.

Bakhu, H.K., 1992. *Herbs that heal*, Orient Paperbacks, Delhi

Chatterjee, T. K. 1997. *Herbal Options*, Eastern Traders, Calcutta.

Chatterjee, A., Pakrashi, S., 1991. *Treatise on Indian Medicinal Plants*, CSIR, New Delhi.

Chakrawarti, H. L., Chakraborty, D. P. 1964. Spices of India, *Indian Agriculturist*, **VIII(2)**, 121.

Cichewicz, R. H., Thorpe, P. A., 1996. The antimicrobial properties of chilli pepper and their uses in Mayan medicine, *Ethnopharmacol*, **52**, 61.

De, A. K., 1994. Chilli, D. K. Publisher, New Delhi.

De, A. K., Ghosh, J. J., 1988. Acute capsaicin treatment causes alteration in rat lung lipids, *Med Sci Res*, **16**, 929.

De, A. K., Ghosh, J. J., 1993. Capsaicin action modulates lipid peroxidation induced by different irritants (review), *Phytother Res*, **7**, 272.

De, A. K., Sengupta, D., Ghosh, J. J., 1992. Capsaicin pretreatment inhibits cigarette smoke induced lipid peroxidation in rat lung. *Med Sci Res*, **20**, 553.

Jain, R. C., 1993 Antibacterial activity of garlic extract. *Ind. J. Med Micro*, **11**, 26.

Khajuria, A., Thusu, N., Zutshi, U., Bedi, K. L., 1997. Antioxidant potential of piperine on oxidant induced alterations in rat intestinal lumen. *Ind Drugs*, **34**, 557.

Khanna, N. M. 1999. Turmeric—Natural precious gift, *Current Science* **76**, 1351

Mehta, R. L., Zayas T.P., Yang, S. S., 1994. Ajowain as a source of natural lipid oxidant, *J. Agri & Food Chem*, **42**, 1420.

Nair, S. C., Verghese, C. D., Paniker, K. P., Kurumboor, S. K., Parathod, R. K., 1994. Effects of saffron on vitamin A levels and its antitumour activity on the growth of solid tumours in mice. *Int J. Pharmacognosy*, **32**, 105.

Piyachaturawat, P., Gausar, R., Suksaram, A., 1996. Choleretic effect of *Curcuma comosa* rhizome extracts in rats, *Pharmacognosy*, **34**: 174.

Raghuram, T. C., Sharma, R. D., Shuvakumar, B., Sahay, B. K., 1994. Effect of fenugreek seeds on intravenous glucose deposition in non insulin dependent diabetic patients. *Phytotherapy Research*, B, B2.

Singh, S., Puri, S., Davi, H. M., Sharma, R., 1997. Studies on the protective properties on garlic oil against acetaminophe induced hepatotoxicity in rat. *Ind J. Pharm. Sci.*, **59**: 259.

Singh, S. S., Agarwal, S. K., Verma, S., Siddiqui, M. S., Kumar, S., 1998. Chemistry of garlic with special reference to allicin and alin–a review. *Ind. Med. & Arom. Plant Sci.*, **20**: 93.

Singh, V. B. and Singh, K., 1996. Spices. Published by New Age International Publishers.

Recent Trends in Spices and Medicinal Plants Research pp. 19-26, 2000
Ed. Amit Krishna De
Associated Publishing Co., New Delhi

1.2

MEDICINAL USES OF SOME SPICE PLANTS AS IN ANCIENT LITERATURE

Anil Kumar Dhiman* and Purshotam Kaushik*

Abstract

The knowledge on the uses of plants as medicines dates back to 4500-1600 BC in the form of hymns in Rigveda which is the earliest known literature in the world. Later, in the three other Vedas, various recipes of plants and their products are met. Works like that of Chakradatta, Sarangdhara and of Bhavprakasha present a more detailed account on various ethnobotanical as well as medicinal uses of different plant species. The spices also possess great medicinal potential, e.g. *Anethum sowa, Brassica juncea, Elatteraia cardamomum, Piper longum* and *Syzygium cumini*. The present paper deals with classic uses of some spice plants, the recipes of which are relevant in present days.

Introduction

The knowledge on the uses of plants as medicines dates back to 4500-1600 BC in the form of scattered hymns in Rigveda which is supposed to be the first known literature on the Earth. Atharvaveda, another important veda from which Ayurveda, the science of life has been derived is known to contain information on about 280 medicinal plants (Kamal, 1988). Besides that, Yajurveda and Samveda also throw some light on different uses of plants.

Kaushik and Dhiman (1997), dealt with the medicinal utilities of some vedic plants. Charak Samhita of Acharya Charak and Susurut Samhita of Susurutcharya are considered major landmarks in the field of Ayurveda and along with Vagbhatta, the writer of Astanga Hridya Samhita are known by the name *Vrihad-trayi* in the Ayurveda. Other works like that of Chakradatta of Chakrapani Datta and Sarangdhara Samhita of Sarangdhara contain lots of information about different uses of plants.

Spice plants apart from serving crucial role in Indian culinary are revealing important properties in preventing or combating certain ailments since ancient times and various references are met in ancient literatures regarding the medicinal uses of these plants and many of them are still useful in modern environment.

*Botany Department, Gurukul Kangri University, Haridwar 249 404.

The present paper deals with such uses of these plants and they are arranged under their common names followed by botanical names and family, besides their medicinal uses with complete references.

Adrak (Ginger) *Zingiber officinale* Rosc. (Zingiberaceae)	Fresh adrak mixed with equal quantity of vinegar is taken in indigestion. (AH U 40.55). 20 gm adrak mixed with equal quantity of jaggery increasing from 20 gm per day upto 200 gm is taken for a month in oedema. After this, diet of rice and soup is to be taken during the course. (CS Ci 12.47-48 and AH Ci 17.6-7) Snuffing of adrak juice is useful in fainting. (BP Ci 1.849). Juice of adrak mixed with honey is taken in cough, anorexia and coryza. (SG 2.1.13). **Dose:** Rhizome juice 5.0-10.0 ml. **Preparations:** Adrak khand, Panchasam churna, Lavanbhaskar churna, Sudarsan churna and Dedarvadi churna
Ajmoda (Celery) *Trachyspermum roxburghianum* (D. C.) Craib. (Umbelliferae)	In urinary stones, ajmoda, gokshuru, talpatri, root-bark of kadamba and bilva and bones of heron, camel and ass together in powder form are taken with water. (AH Ci 11.28-29). In diarrohoea, milk mixed with ghee, honey and powder of sugar, arlu, madhuka and ajmoda is given. (SS U 40.79). Besides, ajmoda, patha, kutaja, utpala, sunthi and pippali in equal quanities are given with tepid water to control the same. (SS U 40.154). **Dose :** Fruit powder 3.0-5.0 gm. **Preparations:** Ajmoda churna, Ajmodadi vatak and Agnitundi vati.
Ajwain (Bishops weed) *Trachyspermum ammi* (L.) Sprague (Syn. *Carum copticum* (L.) C.I. (Umbelliferae)	As an appetizer and carminative, powder of ajwain with butter milk and bida salt is taken. (CS Ci 5.168). Ajwain with sunthi, patha, pomegranate juice and jaggery mixed with salted butter milk is taken as carminative. (CS Ci 14.99). In urticaria, ajwain with jaggery is taken for a week. (VM 52.4). **Dose :** Seed powder 1.0-3.0 gm. **Preparations :** Yavankadi kwath, Yavankadi churna, Yavanisadav and Yavanyark.

Anar (Pomegranate) *Punica granatum* L. (Punicaceae)	Fresh rind of anar fruits with goat's milk is applied on freckles etc. (AS U 37.24). A paste of sarkara and anar or draksha or anar itself is kept in mouth in excessive thirst. (BP Ci. 1.321). A paste of bark of anar plant is applied on soft chancre. (BP Ci 51.26). **Dose :** Bark powder 3.0-5.0 gm and fruit rind powder 3.0-5.0 gm. **Preparations :** Dadima chatusham, Dadimastak, Dadima churna, Dadimadya taila and Dadima ghrita.
Bach (Sweet Flag) *Acorus calamus* L. (Araceae)	Bach with milk or ghee or oil for a month is used as a rasayan. (AH U 39.164). Milk added with water and bach is taken in retention of urine to check upward movement of *vayu.* (BP Ci 31.24) Water boiled with bach and musta and parpata or haridra and sunthi is taken in diarrhoea. (AH Ci 9.8 and CS Ci 19.22). A snuff of bach and pippali with honey is used in headache. (SS U 26.33) **Dose :** Rhizome powder 2.5-5.0 gm. **Preparations :** Bach yog, Medhya rasayana, Yograj guggulu and Chandraparbha vati.
Badi Elaichi (Big Cardamom) *Amomum subulatum* Roxb. (Zingiberaceae)	In equal parts with pippali, it is taken with curd in cough. (SS U 52.19). A decoction of badi elaichi, pippali, sariva, draksha and satpushpa mixed with jaggery is taken to alleviate fever caused by vata. (SS U 39.169 & 171). **Dose :** Seed powder 1.0-3.0 gm. **Preparations :** Mahakalyana ghrita.
Choti Elaichi (Small Cardamom) *Eletteria cardamomum* Maton (Zingiberaceae)	It is taken with wine or amalka juice or mixed with honey along with wine or mixed with honey along with kadali leaf stalks is given in dysuria. (AS Ci 13.5 & AH Ci 11.10). Powder of the seeds dipped in goat's urine for three days is used as a collyrium (CS Ci 26.249). Besides, powder of elaichi, camphor, tagara, kustha and patha mixed with ample salt and oil is

rubbed on wound to induce flow of blood. (SS SU 14.35).

Dose : Seeds 0.5-1.0 gm.

Preparations: Eladi churna, Eladi arista, Eladi gutika, Eladi modak and Eladi kwath.

Dalchini (Cinnamon)
Cinnamomum zeylanicum Blume (Lauraceae)

The bark of the plant with ela, pippali, sarkara and twaksiri successively in double quantities are powdered and taken as linctus to alleviate cough, chest pain and asthma. (AH Ci 5.33-34)

In headache, dalchini, patra and sarkara pounded with rice water is used as snuff followed by ghee snuffing. (CS Ci 26.178)

Dose : Bark powder 1.0-3.0 gm.

Preparations: Samsarkara churna, Sitopaladi churna, Lavanbhaskar churna, Talisadi churna and Chandraparbha vati.

Dhania (Coriander)
Coriandrum sativum L. (Umbellifereae)

Cold infusion of dhania with sugar alleviates burning sensation in fever. (BP Ci 1.356).

In diarrohoea, water processed with dhania and balaka is taken. (BP Ci 2.14).

A decoction of dhania and sunthi is used as digestive and appetizer. (SG 2.2.65).

Dose: 3.0-5.0 gm.

Preparations: Dhanyakadi heem, Dhanyapanchak kwath and Dhanya chatusk.

Haldi (Turmeric)
Curcuma longa L. (Zingiberaceae)

A collyrium made of haldi, red ochre and amalaka is used to alleviate jaundice. (AH Ci 16.44).

In prameha, powder of haldi mixed with honey and juice of amalaka is taken. (AS Ci 14.5; AH Ci 12.5 & CS Ci 6.26).

A piece of haldi and imli leaves are taken with water to alleviate pox (CD 54.9).

In vatarakta diseases, a decoction of haldi and guduchi mixed with honey is taken. (BP Ci 29.79).

A paste of haldi powder mixed with latex of sunthi or pippali and pounded with oxbile is applied in piles. (CS Ci 14.52).

In anaemia, haldi mixed with triphala, ghee and honey is taken. (SS U 44.17).

Dose: Tuberous root juice 5.0-10.0 ml and powder 1.0-3.0 gm.

Preparations: Haridra khand, Punarnava mandur, Chandraparbha vati and Vidangadi lep.

Imli (Tamarind)
Tamarindus indica L.
(Caesalpiniaceae)

In swellings caused by injury, a paste of wheat flour, salt and leaf juice of imli is applied. (BP Ci 48.19).

A hot paste of fruits of imli mixed with sour gruel and oil relieves pain and swelling of fracture. (BP Ci 48.32).

A vegetable of the leaves of plant fried with oil, cooked in curd and added with dhania and sunthi is useful in piles. (CS Ci 14.124-35).

Dose : Fruit pulp 5.0-10.0 gm.

Preparations: Astanga lavan.

Jafran (Saffron)
Crocos sativus L.
(Iridaceae)

As a paste, it is applied in cold. (AS Ci 2.87).

A snuff of ghee cooked with sarkara and jafran relieves headache. (SG 3.3.32).

10 gm of jafran which kept over a night in honey water in the morning is taken in suppression of urine. (SS U 58.31)

Dose: Stigma 0.5-1.0 gm.

Preparations: Kumkumadi ghrita and Kumkumadi taila.

Jaiphal (Mace)
Myristica fragrans Houtt.
(Myristicaceae)

A paste of the arils of jaiphal is used to remove freckles. (BP Ci 61.42).

Water boiled with jaiphal or lavanga and cooled is taken in case of thrist and slight nausea (CD 6.91).

Dose: Aril powder 1.0-2.0 gm.

Preparations: Jaiphaladi churna, Jaiphaladi vati and Jaiphaladi kwath.

Jira (Cumin)
Cuminum cyminum L.
(Umbellifereae)

A powder of jira with jaggery and honey is given in malarial fever. (AS Ci 2.93).

Dose: Fruit powder 3.0-5.0 gm.

Preparations: Jirakadi modar, Jirakadya taila and Jirakadya arista.

Lasun (Garlic)
Allium sativum L.
(Liliaceae)

150 gm dehusked and dried lasun are boiled four times in milk and water to reduce milk and this is used to alleviate gulma diseases. (AH Ci 14.45-46 and CS Ci 5.94-95).

In amavata, decoction of lasun, sunthi and ingudi is taken. (BP Ci 26.39). Besides, 40 gm lasun mixed with 80 gm ghee and a bit honey are taken with milk for a year as a rasayana. (AS U 49.101-34 and AH U 39.111-30).

In malarial fever, lasun mixed with oil is taken before meal. (CS Ci 30.304) or mixed with ghee at every morning. (BP Ci 1.752).

Dose: Tuber paste 3.0-5.0 gm.

Preparations: Rason vati, Rasonpind, Rasonastak and Lasunsdya ghrita.

Lounga (Clove)
Syzygium aromaticum (L.) Merr. & Pers. (Myrtaceae)

As mouth refreshner, it is kept in mouth along with jaiphala, latakasturi, kakkola, tambula, camphor and ela. (CS SU 5.76-77).

Dose: Flower bud powder 1.0-3.0 gm.

Preparations: Lavangadi churna, Lavangachattusa, Lavangadi Vati and Lavangodak.

Marich (Black Pepper)
Piper nigrum L.
(Piperaceae)

Marich is rubbed with curd to form collyrium which is used to cure night blindness. (AH U 13.34)

It is also taken with honey in cough. (AS Ci 4.57-58 and SS U 52.21).

Preparations: Marichadi churna

Dose : Fruit powder 0.5-1.0 gm.

Methi (Fenugreek)
Trigonella foenum-raeceum L.
(Papilionaceae)

Chaturbija, (the seeds of four methi, chandrasura, kalajaji and yavani) powder is taken to alleviate vata diseases, colic, indigestion, flatulance and pain in lumbago. (BPN 1.98-99)

Dose : Seed powder 3.0-5.0 gm.

Preparations : Panchjiraka pak, Methi modak and Swalapa methi modak.

Pippali (Long Pepper)
Piper longum L.
(Piperaceae)

In constipation, a gruel of barley, pippali and amalaka is fried with ghee and given to excrete impurities. (AH Ci 1.31).

Pippali and amalaka powder impregnated with amalaka juice and mixed with sugar, honey and ghee is taken with milk as rasayana. (AH U 40.27)

In fever, pippali with equal quantity of triphala is given and with honey or ghee in cough and asthma. (BP Ci 1.378).

Pippali with honey, butter milk and chitraka is eaten to cure diarrhoea. (CS Ci 19.113 and AH Ci 9.107).

Pippali powder, ghee and honey are added in boiled milk and this milk is used to alleviate cough and fever. (CS Ci 11.79).

Dose : Fruit powder 1.0-2.0 gm.

Preparations : Pippalikhand, Pippalasav and Pippalimuladi kwath.

Piyaz (Onion)
Allium cepa L.
(Alliaceae)

It alone or mixed with meat soup and gruel or khada is used to check bleeding. (CS Ci 14.208)

Roots of piyaz mixed with women's breast milk are put into nostrills to check asthma and hiccough. (CS Ci 17.131).

Dose: Tuberous root juice 5.0-10.-0 gm.

Preparations: Majun piyaz.

Sarson (Mustard)
Brassica campastris L. *var sarson* Prain
(Brassicaceae)

A paste of sarson alone is applied on head in coryza. (AH U 20.13).

The paste of sarson is also applied in oedema. (CS Ci 12.98).

A paste of yellow sarson, tila and asvagandha; priyala, slesmataka and kapittha; madhusigru and punarnava; trikatu, katuka, prisniparni and brathi—these five pastes warm and pounded with alkaline water are used in vatrakta diseases. (SS Ci 5.19).

Sarson mixed with warm water or triphala and trikatu in equal quantities then mixed with bijapura, saindhava and hingu is taken with sour gruel in cholera. (SS U 56.15).

Dose : Seed powder 1.0-3.0 gm.

Preparations: Sarsapadi pralep, karanjadi yog and Vidangadi lep.

Shiajira (Black Cumin)
Carum carvi L. (Apiaceae)

In equal quantities shiajira and jaggery is taken with a bit of maricha in malaria. (BP Ci 1.755).

In anorexia, shiajira, marich, jiraka, draksha, Vrksmala, dadima and suvarchala mixed with jaggery and honey is given. (CS Ci 26.212).

Dose : Fruit powder 1.0-3.0 gm.

Preparations: Hinguvastak churna, Lavanbhaskar churna, Yograj guggulu and Vidangadi louha.

Tejpat (Indian Cassia) *Cinnamomum tamala* Nees & Ebrem (Lauraceae)	Leaves of the plant with some other ingredients are used to cure asthma and cough. (CS Ci 11.21-24). **Dose :** Leaf powder 1.0-3.0 gm. **Preparations:** Eladi gutika, Lavanbhaskar churna and Chandrparbha vati.

Abbreviations

AH = Astand Hardya; AS =Astang Sangrha; BP = Bhavprakash; BPN = Bhavprakash Nighantu; CD= Chakradatta; Ci = Chikitsa Sthana; CS = Charak Samhita; SG = Sarangdhara Samhita; SS = Susrut Samhita; SU = Sutra Sthana; VM = Vrndamadhava.

Conclusion

Thus, we can see that these spice plants are not only used as a spice or condiments but also possess medicinal potential and moreover, most of the above mentioned remedies as recorded in ancient Indian literature, are still relevant in the present environment and being used. The essence of pods of *Vanilla* an orchid (Kaushik, 1983) are used to provide flavour to ice-cream and other food items. The cultivation and some other aspects of some of them are available (Kaushik, 1988). But the need is to popularize the herbal based formulations. It is a proven fact that herbal based remedies do not bear side effects even if taken for a long time. However, it is advised to take these under supervision of a qualified physician to obtain maximum benefits. So, efforts should be made to popularize them and trigger their cultivation.

References

Arundutta, 1925. *Astanga Hridaya Samhita*. Nirnay Sagar Press, Bombay.

Atridev, 1988. *Susrut Samhita*. Motilal Banarsidass, Delhi.

Bennet, S. S. R., 1987. Name Changes in Flowering Plants of India and Adjacent Regions. Triseas Publishers, Dehradun.

Chhangani, G. S., 1945. *Astanga Sangrah*.Chowkhamba Sanskrit Series, Varanasi.

Chunekar, K. C., 1988. *Bhavprakash Nighantu*. Chaukhamba Bharti Academy, Varanasi.

Gupta, A. D., 1951. *Astanga Sangrah Samhita*. Nirnay Sagar Press, Bombay.

Kamal, R., 1988. Economy of Plants in the Vedas. Commonwealth Publishers, New Delhi.

Kanthdutt, 1943. *Vrinda Madhava*. Anand Ashram, Pune.

Kaushik, P., 1983. Ecological and Anatomical Marvels of the Himalayan Orchids. Today and Tomorrow's Printers & Publishers, New Delhi.

Kaushik, P., 1988. Indigenous Medicinal Plants including Microbes and Fungi. Today and Tomorrow's Printers & Publishers, New Delhi.

Kaushik, P. and Dhiman, A. K., 1997. Some Vedic Medicinal Plants. *Ad. Plant Sci.* **9** (II) Suppl.: 1-12.

Shastri, K. N., 1994. *Charak Samhita*. Edited by S. Pandeya. Pt. I & II.Chaukhamba Sanskrit Sansthan, Varanasi.

Srivastava, S., 1996. *Sharangdhara Samhita*. Chaukhamba Orientalia, Varanasi.

Tripathi, J. P., 1961. *Chakradatta*. Edited by B. S. Shastri. Chaukhamba Sanskrit Series Office, Varanasi

Vaish, S. G., 1907. *Bhavprakash*, Khem Raj Shree Krishan Das, Bombay.

Recent Trends in Spices and Medicinal Plants Research pp. 27-30, 2000
Ed. Amit Krishna De
Associated Publishing Co., New Delhi

1.3

ANCIENT USE OF SPICES

Goutam K. Dutta* **and Pratip K. Debnath***

Introduction

Ancient Indian literature is a golden treasure of informatics including the problems of health and diseases, preventing promotion and care, social, cultural, spiritual and religious, climatics, seasonal and environmental aspects including economics. One such important antiquity is the vedic literature, which are four in number, besides the ayurvedic treatise like Charaka Sushrita, Vhagbhat (Asthanga Hridaya) Samhita and Bhavprakash. In the development and evolution of Indian medical sciences (ayurveda) those information formed the basis. In ancient India, sanitation, hygiene etc. in town planning were ranked high and they are forerunners in today's architectural planning with similar spirit as a preventive measure. Health and the maintenance of healthier life style for longivity were the important agenda encircled within the framework of socio-cultural, spiritual and religious aspects to keep the body and mind echoing through ages. Natural elements were the only things to be consumed either as food or as medicines. Besides medicinal plants, spices were of general interest to increase the quality of food and to decrease body ailments.

In medieval India, besides fine textiles and ivory, spices were considered to be highly important. Even today, India is considered as the home of "spices". In ancient economics, spices ranked with precious inventories of royal possessions. It also played an important role in ancient medicine. Even today, spices are used as medicines and playing a significant role in the development of national economy.

Spices in Ayurveda

The vedic literature (2500 BC) is the source of information that contributes to the development of Ayurveda where different parts of the plants have been mentioned as medicines.

NUMBER OF PLANTS UNDER REFERENCE

Rigveda	67	Plants
Samaveda	67	Plants
Yajurveda	100	Plants
Atharvaveda	290	Plants
Brahmana	129	Plants
Upanisada	31	Plants

*J. B. Roy State Ayurvedic College & Hospital, Calcutta.

In the list given here, there are common plants with same nomenclature. Only 16 plants are common to all including *Brahmana* and *Upanisada*. They are *Aswatha*, *Uddojas*, *Khadira*, *Parna*, *Puskara*, *Bivitaka*, *Bilva*, *Madhuna*, *Monja*, *Java*, *Virana*, *Venu*, *Vetasa*, *Shara*, *Soma*, *Somavati*. Proper taxonomical identification of these plants have not yet been done.

In Ayurveda, formulations are found that for the treatment of gastrointestinal tract disorders, especially for the peptic ulcer syndrome or in liver damage, spices contribute a major amount. Though ginger and pepper are pungent in taste, yet their inclusion is inevitable. To Indians, spices will not cause dyspepsia and flatulence due to environmental temperature. In Western style dishes spices are not preferred. To control spoilage of cooked foods, caused by high temperature, addition of spices are desirable. Ginger and pepper (both black and long) are good for peptic ulcer and liver disorder. In general, spices are used in perfumes, offerings, medicine, preservatives, flavours, cultural marks, cosmetics, insecticides, as dyes and condiments.

Throughout the world, pepper is regarded as 'Black gold of India' and 'King of Spices' while cardamom as the "Queen of spices". The medicinal value and therapeutic utility of commonly used spices are mentioned below.

1. Chewable spices—cardamom (small and big), clove funnel, coriander, ajowain, saffron (rarely), pepper mint.
2. Patient's diet—Turmeric, black pepper, cumin, ginger, coriander.
3. Antispasmodic—long pepper, ginger, ajowain, thyme, cumin, turmeric, asafeotida, cinnamon.
4. Antiseptic—garlic, anise, cardamom, clove, cinnamon, cassia, turmeric, ajowain, asafoetida.
5. Antidiabetic—fenugreek, saffron, long pepper, black pepper, turmeric, tejpat, cinnamon.
6. Antihypercholesterimic—capsicum, turmeric, fenugreek, pepper (black pepper) asafoetida, garlic, onion.
7. Lung disorder—asafoetida, tejpat, chilli, long pepper, saffron, mace, turmeric, calamus.
8. Liver disorder—pepper (Long & Black), turmeric, ajowain, fenugreek, coriander, cardamon, cinnamon, fennel, cumin and ginger.
9. Reproductive organs—cinnamon, fenugreek, saffron, mace, nutmeg, turmeric, onion.
10. Immuno-regulator—saffron, turmeric, garlic.

Besides these common uses, spices are included for the treatment of rheumatism, nerve disorders, expectorant, asthma, aphrodisiac, and as stimulant. Presently, spices containing perfume and cosmetics have become popular to the westerners. Natural colour prepared from red chilli have also gained importance for export.

Nomenclature of Plants and Spices

In ancient times, the classification of plants or spices was done in quite a different manner than the present taxonomical classification. In Charaka and Susruta Samhita, the therapeutic classifications are mentioned like Badanahara (analgesics), Dipaniya (digestive) Arshaghna (antihaemorroids) etc.

The nomenclature of plants and spices in Ayurveda depend on any of the following criterion.

1. Utpattibheda—According to area where grown including type of propagation.
2. Sangattanabheda—The parts of the plants are in use
3. Prayogabheda—Pharmacological response
4. Roganusara—According to therapeutic uses in diseases
5. Guna-karma bheda—Properties and qualities of pharmaco—dynamic action.
6. Swarupabheda (Aakriti)—Shape and size
7. Varnabheda —According to colour

In ancient literature (including Ayurveda) different synonyms of a plant have been mentioned. There is every possibility that in ancient times the teaching was "guru-shisya parampara" meaning solo residential schooling system. The nomenclature was different in different schools. However, later, when communication and orientations developed there has been attempts to resolve the controversies and all the names of a plant have been enlisted together. Sometimes, in vedic literature a background story is available in consideration to a name of a plant. The names given to the spices are shown in Table 1.

TABLE 1: SPICES AND THE SYNONYMS GIVEN TO THEM

Common name	*Synonyms*
Maricha	Vellaka, Krishna, Ushna, Dharmapattana
Pipul	Pippali, magadhi, krishnaa, baidchi, chapala, kana upakulya. ushanaa, kola, tikshnatandula
Ada	Katubhadra, sringabera, adraka, adrika
Jamani	Jabanika, ugragandha, brahmadaiva, dipya, dipyaka
Jira	Jiraka, Jrana, Ajajee, Kana
Kalajira	Sugandha, udgarasodhana
Hingu	Ashasryabedhi, Jatuka, ramatta, bannihika
Vacha	Ugragandha, sadagrantha, golomi, sataparbika, kaudripatri, mangalya, jatila, uggra, lomasa.
Lasun	Lahasuna, rasuna, ugragandha, mahaousadhi, arista, rasenaka, jabanesta
Piyaj	Palandu, jabanesta durgandha, mukhadusaka
Jatiphal	Jatiphala, jatikosha, malatiphala
Labanga	Debakusuma, sriprasanuka, srisanga
Elachi (big)	Ela, sthula, bahula, prithika, tripatu, bhadriyala, brihadela, chandrabala
Elachi (small)	Sukhya, upakunchika, tuthya, korangi, dravidaie, tratruti.
Kumkum	Shushrina, rakta, kashmiri, pitaka, bava, senkocha pishuna, dhira, balhika

Tejpata	Gandhajata, patraka, paharanjana, patra, tamalpatra
Daruchini	Guratwaka, twakapatra, bhiringa, chocha, utkrista, twacha
Mouri	Satapuspa, madhura karabi, misi, atichatra, sahitachatrika
Methi	Methini, dipana, bahupatrika, bedhani, gardhabeeja, chandrika, jyoti, gandhaphala, ballari, mantha, mishrapuspa, pitabeeja
Sarisa	Karsyapa, katuka, sneha, tantuva, kadambaja
Posto	Khasbeeja, khakastila
Haridra	Kanchani, pita, krimighni, harabilarini, sositapriya
Tetul	Amlika, chukrika, amli, tintri, chinchika

References

Chopra, R. N., *Glossary of Indian Medicinal Plants*. C. S. I. R., New Delhi.

Ghosal, S. N. *et al.*, 1989. Mangierocin, an adaptogenic Xanthone Carotenoid glycosidic conjugate from saffron. *J. Chem. Res.* 70-71.

Hall, R. L., 1981. The History, Use and Pharmacology of Spices. *Perfumer and Flavorist*, Allured Publishing Corporation, U. S. A, **6**: 1–11.

J. S. Pruthi, *Spices and Condiments*. National Book Trust of India, New Delhi.

Kali Pada Biswas, *Bharater Banausadhi* (1-6 vol); Calcutta Univ. Publication (Bengali).

Kaviraj Shil Kali Bhattacharya,*Chiranjib Banausadhi* (1-12 Vol), Ananda Publishers, Calcutta (Bengali)

Medicinal Plants of India (1-2 vol). ed., G. V. Satyabasi, I. C. M. R. Publication, New Delhi.

Mahendra Kumar Shastri, *Brihat Dravyagunadarsha*. Ayurveda and Tibb Academy U. P., Lucknow (Hindi).

Sriniketan Chakrabarty, *Dravyaguna Samhita*, Ayurveda Niketan, Dhaka, Bangladesh (Bengali).

Sudhendu Das Sharma, *Gharoya Chikitsa*, Asia Publishing Co., Calcutta (Bengali).

The National Formularies (1906-1980) American Pharmaceutical Association, Washington DC (1st-14th ed.) United States Pharmacoepial Convention, Inc, Rockville, Maryland (15th ed.), 1980.

Upendra Nath Sen and Debendra Nath Sen. Dravyaguna, Jabakusum House, Calcutta (Bengali).

Wealth of India, ed., B. N. Sastri, C. S. I. R. New Delhi.

Recent Trends in Spices and Medicinal Plants Research pp. 31-38, 2000
Ed. Amit Krishna De
Associated Publishing Co., New Delhi

1.4

SPICES—VALUED FOR MORE THAN TASTE AND FLAVOUR OF FOODS

K. Srinivasan*

Abstract

Pioneering research during the last three decades have documented many beneficial physiological effects of spices which suggest that the use of spices extend beyond taste and flavour. Digestive stimulant action, hypolipidemic/hypocholesterolemic effect, hypoglycemic potential, prevention and regression of cholesterol gall stones, amelioration of diabetic nephropathy, antioxidant property, anti-inflammatory and anti-arthritic property, anti-mutagenic and anti-carcinogenic property form the list of multi-faceted beneficial influences of spices so far documented. Many of these physiological attributes of spices have the potential of a possible therapeutic application in a variety of disease conditions.

Introduction

Spices are extensively consumed in many parts of the world as food adjuncts to enhance sensory quality of foods. These spice ingredients impart characteristic flavour, aroma or piquancy and colour to foods. Some spices like fenugreek can also modify the texture of food. It is the common experience that their distinct aroma stimulates the appetite.

Not only are these spices used as flavourings and seasonings but they are also used in perfumery, cosmetics, and toiletries. In addition, spices are also recognised to possess a few common medicinal properties—tonic, carminative, digestive stimulant, stomachic, antispasmodic and antihelmenthic (Nadkarni & Nadkarni 1976). Though the scientific basis for such properties are lacking, these undoubtedly efficaceous attributes have made spices find pharmacological application in the indigenous system of medicine in India.

India is the largest producer and consumer of spices. The spice, probably, is the most ancient trade practised by man. The affluence generated by the spice trade has been responsible for several historic voyages and discoveries of new lands. Today, the annual global spice trade is estimated to be in the order of $ 2000

* Central Food Technological Research Institute, Mysore -570 013

million involving a quantity of 500,000 tonnes. India's annual spice export amounts to 1.7 lakh tonnes and of a value of Rs. 600 crores. Pepper (also called 'king of spices') is the major spice commodity exported from India followed by chillies, turmeric, ginger, cardamom, seed spices, curry powders, spice oils and oleoresins.

Spices are used not only individually, but also in the form of mixtures known as 'curry powders' to suit different tastes and dishes. The component(s) of spices responsible for the quality attributes have been designated as 'Active principles', and in many instances they are also responsible for the beneficial physiological effects of spices. The structures of a few well recognised spice active principles are given in Figure 1.

In recent years, many beneficial physiological attributes of spices have been recognised as a result of conserted research efforts by a few research teams in

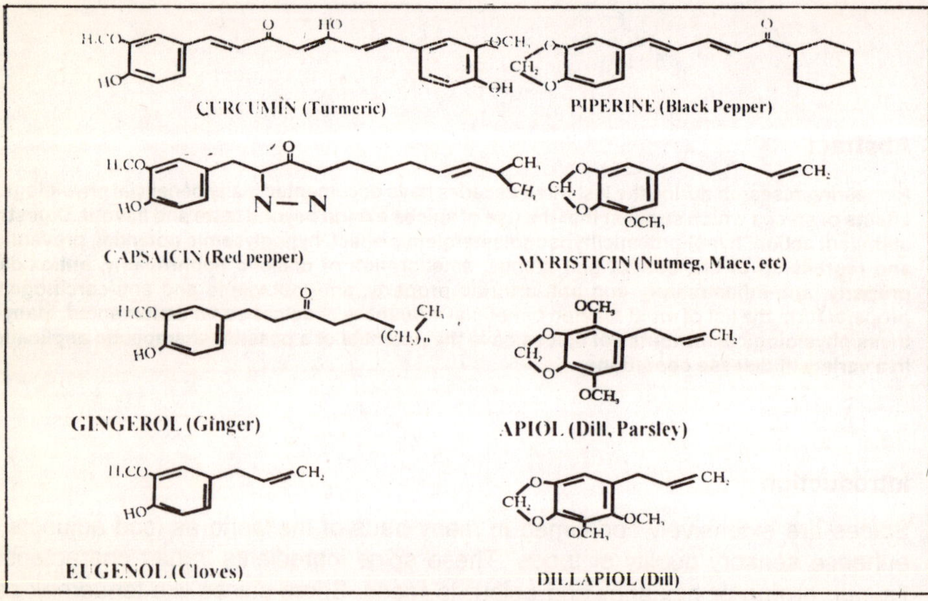

Fig. 1. Active Principle (s) of Spices

India and abroad. (Table 1, 2, 3) Considerable volume of information in this direction has been contributed by CFTRI, Mysore, while NIN, Hyderabad and Kerala University. Trivandrum are other prominent contributers to this aspect. The salient features of a variety of beneficial physiological effects of common spices or their active principles so far documented are summarised in the following paragraphs.

Hypolipidemic/Hypocholesterolemic Effects

The importance of serum cholesterol levels and of lipoproteins in relation to atherosclerosis and coronary heart disease is well known. In view of this, there has been a continuous search for blood cholesterol lowering agents, especially a dietary adjuncts it would be most advantageous. Several commonly used spices have been evaluated for a possible hypocholester-olemic action in a variety of experimental situations in both animals and humans. The spices fenugreek, ginger, red pepper, turmeric, onion and garlic are found to be effective as hypocholesterolemics under conditions of experimentally induced hypercholesterolemia/hyperlipemia. (Singhal *et al.* 1982, Sharma 1986, Jain and Castro 1994; Srimal 1997, Govindarajan and Satyanarayana 1991, Surch and Lee 1995). Further, fenugreek, onion and garlic are effective in humans with hyperlipemic condition. Curcumin and capsaicin, the active principles of turmeric and red pepper respectively are also efficaceous at doses comparable to calculated human daily intake.

TABLE 1: MEDICINAL PROPERTIES OF SPICES RECOGNISED FOR MANY YEARS

Spice	*Medicinal Properties*
Turmeric (*Curcuma longa*)	Anti-inflammatory, diuretic, laxative, good for affections of the liver, jaundice, diseases of blood
Red pepper (*Capsicum annum*)	Anti-inflammatory, for pain relief (Rheumatism and neuralgia), useful in indigestion, rubefacient
Garlic (*Allium sativum*)	Anti-dyspeptic, anti-flatulant, for ear infection, duodenal ulcers, as rubefacient in skin diseases.
Onion (*Allium cepa*)	Diuretic, emmenagogue, expectorant, for bleeding piles
Fenugreek (*Trigonella foenumgraecum*)	Diuretic, emmenagogue, emollient, useful in heart diseases
Cumin (*Cuminum cymminum*)	Antispasmodic, carminative, digestive stimulant
Coriander (*Coriandrum sativum*)	Anti-dyspeptic, flavourant

TABLE 2: HEALTH BENEFICIAL EFFECTS OF SPICES EXPERIMENTALLY DOCUMENTED

S. N.	*Health beneficial effect*	*Spices known to exert*
1	Blood cholesterol lowering	Garlic, Onion, Fenugreek, Turmeric/Curcumin, Chillies/ Capsaicin
2	Prevention and dissolution of cholesterol gallstones	Curcumin, Capsaicin
3	Hypoglycemic	Fenugreek, Garlic, Onion, Turmeric/Curcumin, Cumin
4	Amelioration of diabetic nephropathy	Curcumin, Onion
5	Antioxidant effect	Turmeric/Curcumin, Capsaicin, Eugenol
6	Anti-inflammatory & Antiarthritic	Turmeric/Curcumin, Capsaicin, Eugenol
7	Anti-mutagenic/Cancer preventive	Turmeric/Curcumin, Garlic, Mustard
8	Digestive stimulant	Curcumin, Capsaicin, Piperine, Ginger, Cumin, Mint, Garlic, Onion, Fenugreek
9	Anti-microbial	Turmeric/Curcumin, Asafoetida, Garlic

Hypoglycemic Potential

As part of the dietary treatment of diabetes, there has been a continuous search for noval antidiabetic drugs from plant sources. Spices—the natural and common food adjuncts have also been examined in this direction. Considerable number of human experiments have also been carried out on this aspect, in addition to experimentally induced animal diabetic models. Fenugreek, turmeric or its active principle curcumin, onion or its active principle allyl propyl disulfide, and garlic have been observed to improve glycemic status in diabetic animals/NIDDM patients (Sharma *et al.*, 1996;

Khosla *et al.*, 1995, Carson 1987, Tank *et al.*,1990). Studies on the hypoglycemic property of onion has been most extensive.

Digestive Stimulant Action

Spices like ginger, mint, ajowain and garlic are used as ingredients of pharmacological preparations for digestive disorders. Extensive animal studies have revealed that curcumin (coloring principle of turmeric), capsaicin (pungent principle of red pepper), ginger, fenugreek, mustard, onion and tamarind stimulate bile acid production by the liver and its secretion into bile. Bile acids play a major role in fat digestion and absorption. Curcumin, capsaicin, piperline (pungent principle of black pepper), ginger and mint have been shown to stimulate pancreatic digestive enzymes—lipase, amylase, trypsin and chymotrypsin, which play a crucial role in food digestion. A few of these spices are also found to enhance the activities of lipase and disaccharidases of intestinal mucosa. Thus, many of the common spices act as digestive stimulants by enhancing bile acid secretion and by stimulating the digestive enzymes of pancreas and small intestine (Ploter and Srinivasan, 1996, 1999).

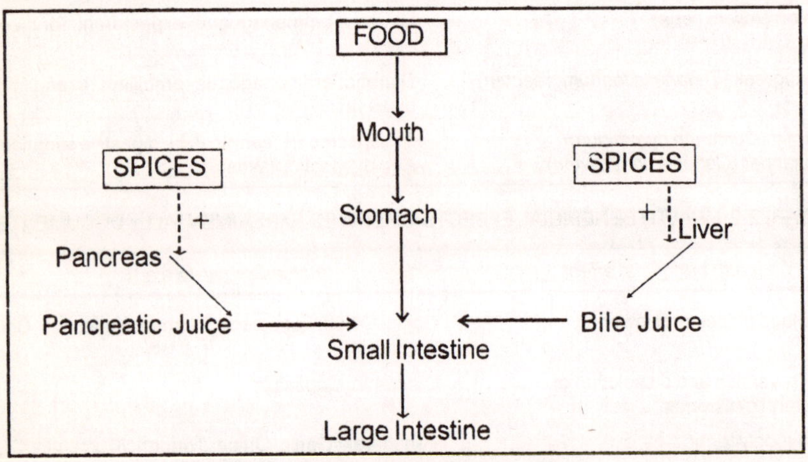

Fig. 2. Digestive Stimulant Action

Prevention and Regression of Cholesterol Gallstone Disease

Studies on experimental induction of cholesterol gallstones in mice and hamsters by feeding a lithogenic diet have revealed that the incidence of gallstones is 40-50% lower when the animals are maintained on 0.5% curcumin or 5 mg% capsaicin containing diet. Animal studies to examine the efficacy of dietary curcumin and capsaicin in regressing the preformed cholesterol gallstones have revealed significant regression by these spice active principles in 10 week feeding trial. The possibility of such a beneficial prevention and regression of cholesterol gallstones by other known hypocholesterolemic spices remains to be examined (Fig. 2 & 3).

Amelioration of Diabetic Nephropathy

High blood cholesterol is an added risk factor that determines the rate of decline of kidney function in diabetic situation. Hypercholesterolemia is a regular biochemical abnormality in diabetes mellitus and nephropathy is one of the secondary eventuality of this disease. Extensive studies with diabetic rat models have been carried out to examine whether hypolipidemic spices supplementation in the diet would influence renal lesions associated with diabetes by virtue of their possible cholesterol lowering effect. Curcumin (active principle of turmeric) and onion have been found to have a promising ameliorating influence on the severity of renal lesions in diabetic rats.

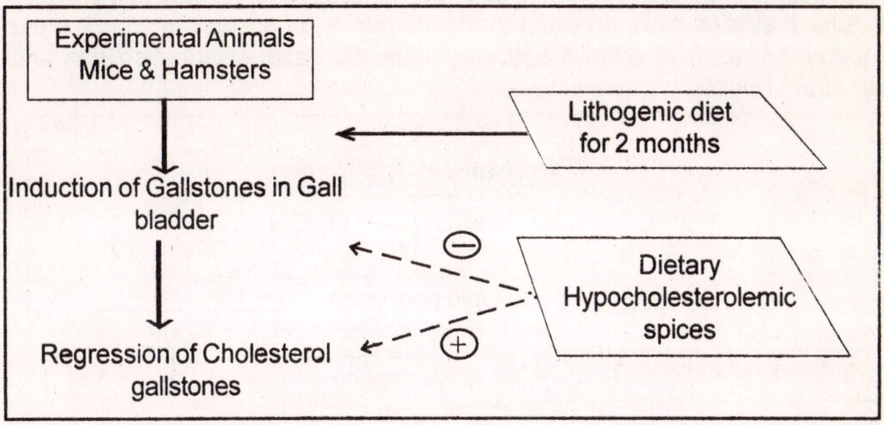

Fig. 3. Dietary Spices & Cholesterol Gallstone Disease

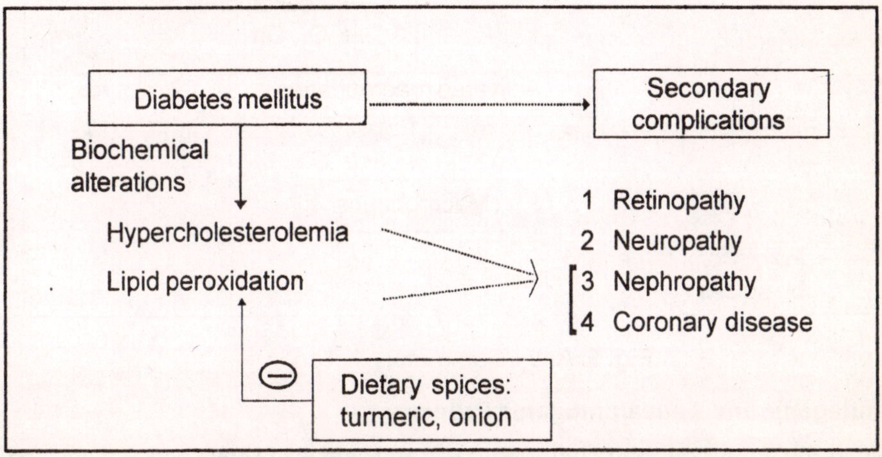

Fig. 4. Amelioration of Diabetic Nephropathy

Antioxidant Property and Ability to Scavenge Reactive Oxygen Radicals

Reactive oxygen radicals are detrimental to cells at both membrane and genetic levels. They induce lipid peroxidation in cellular membranes, generating lipid peroxides which cause extensive damage to membranes, in terms of cross linking of membrane components, leaks and lysis and membrane-mediated chromosomal damage. Studies with *in vitro* systems have revealed that spice principles curcumin (turmeric), eugenol (cloves) and capsaicin (red pepper) have documented beneficial antioxidant property, spice principles—curcumin and eugenol, spices—garlic and asafoetida are shown to have the ability to scavenge superoxide anion and hydroxyl radical *in vitro* systems. (Reddy and Lokesh, 1992; 1994).

Anti-inflammatory and Antiarthritic Property

Both *in vitro* and *in vivo* animal experiments have documented the anti-inflammatory potential of spice principles curcumin, capsaicin and eugenol (Reddy and Lokesh, 1994). Animal studies have revealed that curcumin and capsaicin also lower the incidence and severity of arthritis and also delay the onset of arthritis (Bina and Lokesh, 1997) (Fig. 5).

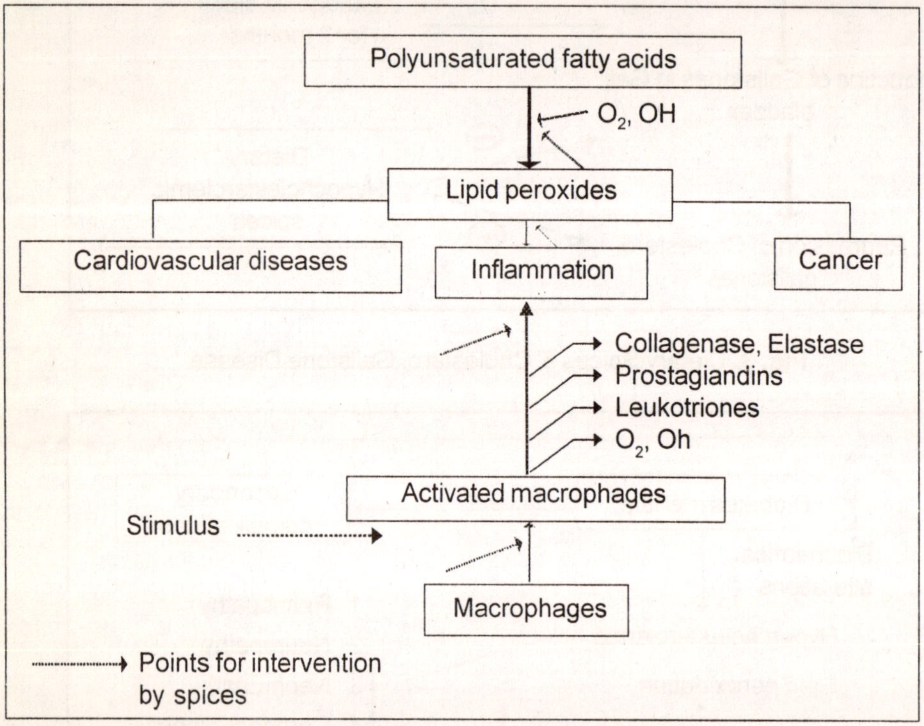

Fig. 5. Anti-inflammatory Action

Antimutagenic and Anticarcinogenic Property

Curcumin, the active principle of turmeric, garlic and asafoetida have been recognised to be antimutagenic. Animal studies involving experimental induction of tumour of specific tissues have revealed significant reduction of the incidence of tumour in curcumin treatment. (Srimal, 1997 Conney *et al.*, 1997)

While the above listed beneficial facets of these common food adjuncts, namely spices, are the outcome of pioneering research over the past two to three decades, there is every scope for further concrete knowledge pouring in over the coming years in this regard.

Antimicrobial Activity

The antibacterial property of turmeric is well known. Its active principle, curcumin is a known bacteriostatic agent whereas the essential oil of turmeric is both antibacterial and fungistatic (Govindarajan 1980). Asafoetida is another spice which is used as an antimicrobial agent in traditional medicine. Its uses in treating chronic bronchitis, whooping cough have been documented in ancient literature. Its use in counteracting intestinal flatulence is also well known. The antibacterial properties of garlic and onion are also long known. The sulphur containing compounds from these plants act against both gram positive and gram negative bacteria (Carson, 1987). The extracts of garlic and onion are known to inhibit growth of many pathogenic fungi belonging to aspergillus, candida and other species (Carson, 1987). Other spices like nutmeg, saffron, cumin, thyme etc., have antimicrobial potential. The volatile component of these spices is believed to be responsible for this property.

Other Beneficial Physiological Effects

Spices have several other beneficial effects which are also important contributors to human health. These include: anti-obesity effect of capsaicin(Govindarajan and Satyanarayana, 1991), anti-cataractogenic effect of curcumin and piperine acting as enhancer of the bioavailability of drugs.

Conclusion

The list of beneficial attributes of these common food adjuncts, spices is increasing with more and more experimental documentation, several of these with proven effectiveness even in human subjects. A few beneficial physiological effects of spices have the potential of a possible therapeutic exploitation in a variety of disease conditions in the years to come. In view of the many promising health beneficial effects spices are understood to exert, these food adjuncts are now considered as the natural and necessary component of our daily nutrition and not as just a food additive to impart taste and flavour to our food.

Reference

Carson, J. F. 1987. Chemistry and biological properties of onion and garlic. *Fd Rev Int,* **3**, 71-103.

Govindarajan, V. S. and Satyanarayana, M. N. 1991. Capsicum-production, technology, chemistry and quality: part 5, Impact on physiology, nutrition and metabolism, structure, pungency, pain and desensitisation sequences. *Cri Rev Fd Sci Nutr.,* **29**, 435-74.

Govindarajan, V. S. 1980. Turmeric—chemistry, technology and quality. *Cri Rev Fd Scvi Nutr.,* **18**, 199-301.

Jain, M. K. and Apitz Castro, R. 1994. Garlic: A matter for heart. In Spices, Herbs and edible Fungi, Ed: Charalambouis G. Elsevier, 311-64.

Khosla, P., Gupta, D. D. and Nagpal, R. K. 1995. Effect of *Trigonella foenumgraecum* on blood glucose in normal and diabetic rats. *Ind. J. Physiol. Pharmacol,* **39**, 173-74.

Nadkarni, K. M. and Nadkarni, A. K. 1976. *Indian Materia Medica*, Popular Prakashan Pvt. Ltd. Mumbai.

Platel, Kalpana and Srinivasan, K. 1996. Influence of dietary spices or their active principles on digestive enzymes of small intestinal mucosa in rats. *Int. J. Fd. Sci. Nutr.,* **47**, 55-59.

Platel. Kalpana and Srinivasan, K. 1999. Influence of dietary spices or their active principles on pancreatic digestive enzymes in albino rats. *Nahrung,* **43**, in press.

Reddy, A. C. P. and Lokesh, B. R. 1992. Studies on spice principles as anti-oxidants in the inhibition of lipid peroxidation of rat liver microsomes. *Mol Cell Biochem,* **111**, 117-24.

Reddy, A. C. P. and Lokesh, B. R. 1994. Alterations in lipid peroxidation in rat liver by dietary n-3 fatty acids: modulation of antioxidant enzymes by curcumin, eugenol and vitamin E. *J. Nutr. Biochem.*, **5**, 181-88.

Bina, Joe and Lokesh, B. R. 1997. Prophylactic and therapeutic effects of n-3 pufa, capsaicin and curcumin on adjuvant induced arthritis in rats. *J. Nutr. Biochem.*, **8**, 397-407.

Sharma, R. D., Sarkar, A., Hazra, D. K., Misra, B., Singh, J. B., Sharma, S. K., Maheshwari, B. B. and Maheshwari, P. K. 1996. Use of fenugreek seed powder in the management of NIDDM. *Nutr. Res.* **16**, 1331-39.

Singhal, P. C., Gupta, R. K. and Joshi, L. D. 1982. Hypocholesterolemic effect of *T. foenumgraecum. Nutr. Rep. Int.*, **33**, 669-77.

Srimal, R. C. 1977. Turmeric: A brief review of medicinal properties. *Fitoterapia,* **LXVIII**, 483-49.

Surh, Y. J. and Lee, S. S. 1995. Capsaicin, a double edged sword: Toxicity, metabolism and chemopreventive potential. *Life Sci.*, **56**, 1845-55.

Tank, R., Sharma, N., Sharma, T. and Dixit, V. P. 1990. Antidiabetic activity of *Curcuma longa* in alloxan induced diabetic rats. *Indian Drugs,* **27**, 587-89.

Recent Trends in Spices and Medicinal Plants Research pp. 39-46, 2000
Ed. Amit Krishna De
Associated Publishing Co., New Delhi

1.5

PRELIMINARY SCREENING FOR ANTIMICROBIAL ACTIVITY OF INDIAN SPICES

Minakshi De*, A. B. Banerjee* and Amit Krishna De**

Abstract

Spices have been used in India since ages, for the preservation of food products. Some spices have been reported to have antiseptic and disinfectant influences on the microbial world and are considered as antibacterial and antifungal. In this respect, a preliminary screening for antimicrobial activities of some Indian spices have been carried out. Of the 28 different spices surveyed till date, results indicate clove, cinnamon, ajowain, chilli, horse raddish, cumin, tamarind, black cumin, pomegranate seeds, nutmeg, garlic, onion, tejpat, cellary, cambodge, etc. have potent antimicrobial activities. The test organisms used were *Bacillus subtilis* [ATCC 6633), *Escherichia coli* [ATCC 10536]. *Saccharomyces cerevisiae* [ATCC 9763]. Investigations are now being carried out to identify the antimicrobial components in each spice and to clarify their mechanism of actions.

Introduction

India is known as the "Kingdom of spices". Spices have been used since ages not only for increasing the flavour of food, for their preservative properties (Ayres *et al.*, 1980) but also because of their immense medicinal properties. General activity of spices include antidiabetic, analgesic, anti-inflammatory, antifertility, antioxidant and anticarcinogenic properties.

Spices have also been reported to prevent microbial spoilage of foods (Curtis *et al.* 1996; Hosono A. 1996; Daw Z. Y. 1994). This preservative property of spices has been attributed to the presence of some antimicrobial principles contained in their oils (Sreenivasamurthy and Krishnamurthy, 1959). Capsaicin from chilli acts as a potent inhibitor of the gastric pathogen, *Helicobacter pylori* [Jones *et al.*, 1997]. Essential oil from spices like cumin, ajowain, ginger, coriander, basil, etc has been found to show various degrees of inhibition against *A. niger, S. cerevisiae, Mycoderma* spp. (Meena and Sheti, 1994). Studies on the fungitoxic properties of essential oil of *Cinnamomum zeylanicum* [R. Tewari, 1994] have been carried out.

* Department of Biochemistry, University College of Science, 35 Ballygunge Circular Road, Calcutta 700 019.
** Indian Science Congress Association, 14, Dr. B. C. Guha St, Calcutta 700 017.

Inhibitory effects of anethole and eugenol on the growth and toxin production of *A. parasiticus* (Karapinar, M., 1990), inhibitory effects of spices like *Alpinia galanga, Curcuma longa*, etc against *S.aureus, S.epidermis, S. haemolyticus, E. coli, K. pneumoniae, Enterobacter agglomerans,* etc. [Thomas *et al.* 1996], are some of the recently published reports regarding the antimicrobial properties of spices.

However, these reports are restricted to mainly a few spices (Web and Tanner, 1945, Anderson *et al.* 1953; Conner and Beachut, 1984). The Indian Spices Board at Cochin recognises 52 spices which are produced, cultivated and exported by the country. Most of them require a detailed study, specially for antimicrobial properties, so that in future they can effectively replace the chemically synthesised antibiotics which have a large number of toxic side effects.

In the present study, an attempt has been made to investigate the antimicrobial properties of Indian spices at both antibacterial and antifungal levels. The present communication deals with the investigation of 28 spices against a gram positive, a gram negative bacteria and a yeast. The list includes spices with already established well pronounced activities in order to make a comparative study with the presently reported active spices and also to standardise the method employed for this screening purpose.

Materials and Methods

Sources of Spices

The spices used for this study were either bought fresh from the local markets or were supplied by the Indian Spice Board, Cochin.

Test Organisms

The test organisms used were *Bacillus subtilis* [ATCC 6633], a gram positive organism, *Escherichia coli* (ATCC10536), a gram negative organism and *Saccharomyces* cerevisiae [ATCC 9763], a yeast. These organisms were bought from the Central Drug Laboratory and were maintained at 4°C after subculturing every two months. Regarding the yeast subculture, the method by Gupta and Banerjee (1970) was followed. The solvents and the chemicals used were all from E. Merck.

Solvent Extraction

The fresh spices were carefully dried in the shade and powdered in a mixer. About 5 gms of this was powdered and the grounded mass was extracted with different solvents. The solvent (other than sterile water) was ethyl alcohol of 90 percent purity. For this extraction purpose, the grounded spice was mixed with the solvent (amount of it is 5 times that of the mass of the spice taken) and placed on a rotary shaker for 12 hours. The process was repeated for several times till the extract was free from colour and odour. It was then cold centrifuged and concentrated to a fixed volume and stored in sterile vials at 10ºC. The pH of the extracts was adjusted to 6.7 in order to eliminate the possible interference of the growth of test organisms due to pH shifts. The involved method is a modified form of serial dilution technique. A series of test tubes were taken, which contained 5 ml of sterile nutrient agar media in the molten stage with its temperature being maintained at 45-50°C. A fixed volume of 0.05 ml extract of spice was added to each of the tubes so that the final concentration of spice varied from 1-100 mg/ml in the tubes. The whole process was done asceptically to avoid contamination and it was repeated several times for each of the spices.

The media was then allowed to solidify in the form of a slant and then they were seeded with the test organisms. Nutrient agar media was used for *Bacillus subtilis* and *Escherichia coli* and yeast media was used for *Saccharomyces cerevisiae.* Control tubes were run simultaneously with identical solvent concentration in order to check their possible growth inhibitory effects. That concentration of the spice at which no growth of the organism was observed was considered as the MIC of the spice for that organism. Results were expressed in terms of growth of organisms (+) or inhibition of growth (–) after addition of the spices. Values are expressed as means of six experiments.

Results and Discussion

The results [Table 1] indicate that cumin, cinnamon, black cumin, clove, onion, ajowain, chilli, garlic, cellary, basil, tejpat, nutmeg, small cardamom, caraway, turmeric, tamarind, aniseed, horse raddish, pomegranate seeds and cambodge have antimicrobial properties. Out of these, clove, ajowain, chilli, horse raddish, pomegranate seeds and cambodge have both antibacterial and antifungal properties, while black cumin, basil, tejpat, small cardamom and aniseed have only antifungal properties.

Very little work has been done till date to establish the antimicrobial properties of spices even though the traditional uses of spices as food preservatives and antiseptics are well known. The present study clearly indicates that out of the 28 spices surveyed till date nearly 20 have antimicrobial properties. Some of these have both antibacterial and antifungal properties while others have either antibacterial or antifungal properties. Some recent reports regarding the antimicrobial properties of spices clearly establishes the results of the present study. Some are mentioned below.

The antimicrobial properties of capsicum species and their uses in Mayan medicines (Cichewicz and Thorpe, 1996), antimicrobial and mutagenicity of allyl isothiocyanate from horse raddish and some other essential oils like eugenol, etc from spices (Kano *et al.* 1995), toxicity of clove eugenol against several pathogenic fungi (Tombe *et al.,* 1995), antimicrobial activity of *Nigella sativa* and *Zingiber officinale,* (Kandil *et al.,* 1994), antimicrobial activity and distilled components of garlic (Ji *et al.,* 1997), studies on the antifungal properties of essential oils of *Trachyspermum ammi* (Tripathy, *et al.,* 1986) have been reported.

From the study of the inhibitory doses, it is also clear that clove, horse raddish, ajowain, chilli, and pomegranate seeds have potent antimicrobial properties and needs to be investigated thoroughly. Some work already done in this regard are inhibition of food borne pathogens by eugenol from clove and thymol from ajowain (Karapinar and Aktug, 1987), antimicrobial activity of Capsaicin from *Capsicum frutescens* (Chowdhuri and De, 1986), inhibitory effects of crushed horse raddish on the growth of microorganisms (Chemical Abstracts, 1939), antibacterial properties of the oil of pomegranate seeds (Heftman *et al.,* 1966).

This is a preliminary report on the study of spices for their antimicrobial properties. The authors are in the process of screening and investigating the other

TABLE 1: EFFECT OF DIFFERENT SPICES ON THE GROWTH OF TEST ORGANISMS

Spices		*Test Organism*		
Name	*Dose (mg/ml)*	*B. Subtilis*	*E. coli*	*S. cerevesia*
1. Cumin	1	+	+	+
[*Cuminum cyminum* L.]	2	+	+	+
	5	+	+	+
	10	–	+	+
	25	–	+	+
	50	–	+	+
	100		+	+
2. Cinnamon	1	+	+	+
[*Cinnamomum zeylanicum* Blume]	2	+	+	+
	5	+	+	+
	10	+	+	+
	25	+	+	+
	50	+	+	+
	100	–	+	+
3. Black cumin	1	+	+	+
[*Nigella sativa* L.]	2	+	+	+
	5	+	+	+
	10	+	+	+
	25	+	+	–
	50	+	+	–
	100	+	+	–
4. Clove	1	+	+	+
[*Syzygium aromaticum*	2	+	+	+
L. Merr. et. Perry]	5	–	+	+
	10	–	+	+
	25	–	–	–
	50	–	–	–
	100	–	–	–
5. Onion	1	+	+	+
[*Allium cepa* L.]	2	+	+	+
	5	+	+	+
	10	+	+	+
	25	+	+	+
	50	+	+	+
	100	–	+	+
6. Ajowain	1	+	+	+
[*Trachyspermum ammi*	2	+	+	+
L. Sprague]	5	+	+	+
	10	–	–	–
	25	–	–	–
	50	–	–	–
	100	–	–	–
7. Chilli	1	+	+	+
[*Capsicum annum* L.]	2	+	+	+
	5	+	+	–
	10	+	+	–
	25	–	+	–
	50	–	–	–
	100	–	–	–

8. Garlic	1	+	+	+
[*Allium sativum* L.]	2	+	+	+
	5	+	+	+
	10	+	+	+
	25	+	+	+
	50	–	+	+
	100	–	+	+
9. Cellary	1	+	+	+
[*Apium graveolens* L.]	2	+	+	+
	5	+	+	+
	10	+	+	+
	25	+	+	+
	50	–	+	+
	100	–	+	+
10. Basil	1	+	+	+
[*Ocimum santum* L.]	2	+	+	+
	5	+	+	+
	10	+	+	+
	25	+	+	+
	50	+	+	–
	100	+	+	–
11. Tejpat	1	+	+	+
[*Cinnamomum tamala* Nees]	2	+	+	+
	5	+	+	+
	10	+	+	+
	25	+	+	+
	50	+	+	–
	100	+	+	–
12. Nutmeg	1	+	+	+
[*Myristica fragrans* Houff.]	2	+	+	+
	5	–	+	+
	10	–	+	+
	25	–	+	+
	50	–	+	+
	100	–	+	+
13. Small Cardamom	1	+	+	+
[*Elletaria cardamomum* Maton]	2	+	+	+
	5	+	+	+
	10	+	+	+
	25	+	+	+
	50	+	+	+
	100	+	+	–
14. Caraway	1	+	+	+
[*Carum carvi* L.]	2	+	+	+
	5	+	+	+
	10	+	+	+
	25	+	+	+
	50	+	+	–
	100	+	+	–
15. Turmeric	1	+	+	+
[*Curcuma longa* L.]	2	+	+	+
	5	+	+	+
	10	+	+	+

	25	–	+	+
	50	–	+	+
	100	–	+	+
16. Tamarind [*Tamarindus indica* L.]	1	+	+	+
	2	+	+	+
	5	+	+	+
	10	+	+	+
	25	–	+	+
	50	–	–	+
	100	–	–	+
17. Aniseed [*Pimpinella anisum* L.]	1	+	+	+
	2	+	+	+
	5	+	+	+
	10	+	+	+
	25	+	+	+
	50	+	+	+
	100	+	+	–
18. Pepper [*Piper nigrum* L.]	1	+	+	+
	2	+	+	+
	5	+	+	+
	10	+	+	+
	25	+	+	+
	50	+	+	+
	100	+	+	+
19. Horse raddish [*Armoracia rusticana*]	1	+	+	+
	2	+	+	+
	5	–	–	+
	10	–	–	+
	25	–	–	+
	50	–	–	–
	100	–	–	–
20. Pomegranate seeds [*Punica granatum* L.]	1	+	+	+
	2	+	+	+
	5	+	+	+
	10	–	+	+
	25	–	+	+
	50	–	+	+
	100	–	–	–
21. Cambodge [*Garcinia cambogia* Gaertn. Desr.]	1	+	+	+
	2	+	+	+
	5	–	+	+
	10	–	+	+
	25	–	+	+
	50	–	+	+
	100	–0	+	–
22. Mustard [*Brassica juncea* L.]	1	+	+	+
	2	+	+	+
	5	+	+	+
	10	+	+	+
	25	+	+	+
	50	+	+	+
	100	+	+	+
23. Coriander [*Coriandrum sativum* L.]	1	+	+	+
	2	+	+	+

	5	+	+	+
	10	+	+	+
	25	+	+	+
	50	+	+	+
	100	+	+	+
24. Fennel	1	+	+	+
[*Foeniculum vulgare* Mill]	2	+	+	+
	5	+	+	+
	10	+	+	+
	25	+	+	+
	50	+	+	+
	100	+	+	+
25. Ginger	1	+	+	+
[*Zingiber officinale* Rosc]	2	+	+	+
	5	+	+	+
	10	+	+	+
	25	+	+	+
	50	+	+	+
	100	+	+	+
26. Large cardamom	1	+	+	+
[*Amomum subulatum* Roxb.]	2	+	+	+
	5	+	+	+
	10	+	+	+
	25	+	+	+
	50	+	+	+
	100	+	+	+
27. Curry leaf	1	+	+	+
[*Murraya koenigii* L.Sprengel]	2	+	+	+
	5	+	+	+
	10	+	+	+
	25	+	+	+
	50	+	+	+
	100	+	+	+
28. Mint	1	+	+	+
[*Mentha piperita* L.]	2	+	+	+
	5	+	+	+
	10	+	+	+
	25	+	+	+
	50	+	+	+
	100	+	+	+

+ indicates growth

– indicates no growth

spices under the category listed by the Spices Board. A detailed report is expected to be published very soon.

Acknowledgement

The authors are indebted to Mr P. S. Thampi, Spices Board, Cochin for providing spices on complimentary basis.

References

Anderson, E. E., Esselen, W. B. and Handleme, N. A. R., 1953. The effect of essential oils on the inhibition and thermal resistance of microorganisms in acid food products. *Food Resistance*. **18** : 40 -47.

Ayres, J. C., Mundt, J. O. and Sandine, W. E., 1980. Microbiology of Foods. H. H. Freeman and Company, San Francisco, USA.

Cichewixz, R. H., Thorpe, R. H., 1996. The antimicrobial properties of Chilli peppers [*Capsicum* species] and their uses in Mayanmedicine. *Journal of Ethnopharmacology,* **52:** 61-70.

Conner, D. E., Beachut, L. R., 1984. Effect of essential oils from plants on the growth of food spoilage yeast, *J.Food Science, **49:*** 429-434.

Chemical Abstract, 1939. Inhibitory effect of allyl isothiocyanate isolated from Horse raddish on the growth of microorganisms.

Chowdhury, B. and De, A. K., 1996. Antimicrobial activity of capsaicin from *Capsicum frutescens. Science & Culture.,* **62** (9 & 10), 255.

Curtis, O. F., Shetty, K., Cassagnol, G., Peleg, M., 1996. Comparison of the inhibitory and lethal effects of synthetic versions of plant metabolites [anethole, eugenol, thymol] on a food spoilage yeast (*Debaromyces hansenii*). *Food Biotechnology.* **10** (1): 55-73.

Daw, Z. Y., 1994. Inhibition of *Aspergillus parasiticus* growth and aflatoxin production by some essential oils. *Chemie Mikrobiologie Technologie der Lebensmittel* **16** [5-6]:129-135.

Gupta, S. and Banerjee, A. B., 1970. A rapid method of screening antifungal antibiotic producing plant. *Ind. J. of Exp. Biol.* **8**: 148-149.

Heftmann, 1996. Antibacterial properties of the oil of *Punica granatum. Phytochemistry.* **5:** 337.

Hosono, A., 1996. Effect of spice extract of fungal inhibition. *Lebensmitte: Wissenschoft and Technologie* **29** [3] : 286-288.

Jain, R. C., 1993. Antibacterial activity of garlic extract. *Indian Journal of Medical Microbiology,* **11** [1] : 26-31.

Ji, W. D., Jeong, M. S., Chung, H. C., Lee S. J., Chang, Y. G., 1997. Antimicrobial activity and distilled components of garlic. *Agricultural Chemistry and Biotechnology,* **40** [6]: 514-518.

Jones, N. L., Shabib, S., and Sherman, P. M., 1997. Capsaicin as an inhibitor of the growth of gastric pathogen *Hellcobacter pylori., Fems Microbiology Letter,* **146** (2) : 223-227.

Karapinar, M., 1990. Inhibitory effects of anethole and eugenol on the growth and toxin production of *A. parasiticus. Int. J. Food Microbiology,* **10** [3-4] : 113-119.

Karapinar, M., Aktug, S. E., 1987. Inhibition of food borne pathogens by thymol, eugenol, menthol and anethole. *Int. J. Food Microbiol,.* **4** [2] : 161-166.

Kandil, O., Radwan, N. M., Hassan, A. B., Amer, A. M. M., El-Banna-H. A., Amer, W. M. M., Amer, M. M., 1994. Antimicrobial Activity of *Nigella sativa* and *Zingiber officinale. Veternary-medica;-Journal* -Giza, **42:** 1(A), 159-168.

Kono, M., Yoshida, Y., Itaya, Shimobok, Y., Yoshikawa, Y., Terrashita, T., 1995. Antimicrobial activity and mutagenicity of allylisothiocyanate and several essential oils from spices. *Memoirs of the Faculty of Agriculture -of - Kinki University.* **28:** 11-19.

Meena, M. R. and Sethi, V. J., 1994. Antimicrobial activity of essential oil from spices. *Food Science Technology,* **31:** 68-71.

Sreenivasamurthy, V. and Krishnamurthy, K., 1959. Place of spices and aromatics in Indian dietary. *Food Science* **8:** 284294.

Tewari, R., 1994. Studies on fungitoxic properties of essential oil of *Cinnamomum zeylanicum* Breyn. *Indian Perfumer,* **38** [3]: 98-104.

Thomas Elsamma, Shanmugam, J. and Mahammas, R., 1966. Antibacterial activities of plants belonging to Zingiberaceae family. *Biomedicine,* **16:** (2 & 3) : 15-20.

Tombe, M., Kabayashi, K., Oniki, M. and Ogoshi, A., 1995. Toxicity of clove, eugenol against several pathogenic fungi. *Indonesian Journal of Crop Science,* **10** [1] :11-18.

Tripathy, S. C., Singh, S. P and Dube, S., 1996. Studies on the antifungal properties of the essential oil of *T. aammi. Phytopathology,* **116** [2]: 113-120.

Web, A. H., and Tanner, F. W., 1945. Effect of spices and flavouring materials on the growth of yeast. *Food Resistance,* **10:** 279-282.

Recent Trends in Spices and Medicinal Plants Research pp. 47-61, 2000
Ed. Amit Krishna De
Associated Publishing Co., New Delhi

1.6

ANTIMICROBIAL ACTIVITIES OF AJOWAIN AND CUMIN

Lopamudra Mukherjee*, Minakshi De*,
A. B. Banerjee* and Amit Krishna De

Abstract

Ajowain (*Trachyspermum ammi* L. Sprague) and cumin (Cuminum cyminum) are the two common ingredients of Indian food. These two spices have been used in traditional medicine for ages for curing gastric disorders, flatulence, diarrhoea dyspepsia, etc. and also as antiseptic agent. Extensive studies in our laboratory have revealed that ajowain and cumin have potent antimicrobial activity against a variety of bacterial and fungal species, both pathogenic and nonpathogenic. Dose used for the study varied from 1 mg/ml to 100 mg/ml of a crude alcoholic extract. The results indicate the possible therapeutic uses of ajowain and cumin as potent antimicrobial agents.

Introduction

Antibiotics are defined as substances (secondary metabolites) of microbial origin that have antimicrobial activity in very small amounts. They are classified into two groups. (1) Broad spectrum and (2) Narrow spectrum antibiotics (Davis, 1980).

Broad spectrum antibiotics have the ability to destroy or inhibit many different species of pathogenic micro-organisms of various kinds. On the contrary, the narrow spectrum antibiotics are those which have the ability to destroy or inhibit only small number of pathogenic and nonpathogenic micro-organisms (Pelczar *et al.*, 1986).

The selection of ajowain and cumin as sources of antimicrobial agent is based on five important reasons viz.

(1) Ajowain and cumin are considered as antimicrobial agents because of their traditional uses as antiseptics, disinfectant and antihelminthics etc. These are well supported from the below mentioned published and relevant literatures. (2) They are produced in India in large quantities. (3) These spices are available at relatively low cost. (4) These spices are consumed with no apparent side effects. (5) They are also very popular as well as widely used in almost all parts of our country.

*Department of Biochemistry, University College of Science, Calcutta -700 019.
*Indian Science Congress Association, 14 Biresh Guha Street, Calcutta -700 017.

Some Important Properties of Cumin and Ajowain

Cumin

The scientific name of cumin is *Cuminum cyminum* (Umbelliferae). It is known as "Jira or Sada Jeera" in Bengali and Hindi, as "jiregire" in Marathi, as "siragam" in Tamil, and as "Zira-safud" in Punjabi.

Description of the Plant

Cumin is produced by small slender annual herb about 1 ft high with a much branched angular or striated stem bearing 2 or 3 partile linear leaves, bluish green in colour. The flowers are white or rose coloured born in compound umbels. The fruits are greyish tapering towards both base and apex and compressed laterally covered by papillose hairs.

It is cultivated in most of the Indian states excepting Bengal and Assam. It is cultivated on liberally manured well drained and rich loamy soil.

Uses

Cumin seeds have an aromatic odour and a spicy but slight bitter taste and are extensively used as condiment. In indigenous medicine, cumin seeds have long been considered as stimulant and carminative. They are stomachic, astringent and useful in diarrhoea and dyspepsia. They are now widely used in veterinary medicine (The Wealth of India, 1950).

Composition of Cumin Seed *(in percentage (%) (Pruthi, 1979)*

Component	*Percentage Composition*
Moisture	6.2
Protien	17.7
Fat	23.8
Crude fibre	9.1
Carbohydrates	35.5
Mineral water	7.7
Calcium (Ca)	09
Phosphorus (P)	0.45
Iron (Fe)	0.048
Sodium (Na)	0.16
Potassium (K)	2.1
Vitamin mg/100gm	*Composition/100 gm.*
Vitamin B_1	0.73
Vitamin B_2	0.38
Niacin	2.5
Vitamin C	17.2
Vitamin A	17.51
Food Energy	460 Cal/100gm

The chief constituents of the volatile oil is Cumaldehyde $C_{10}H_{12}O$ (P-isopropyl benzaldehyde, b. p. 235 °C) which forms nearly 22% to 40% of the oil. This oil contains p-cymene, pinene, dipentene, cumene, cuminic alcohol, β-phellandrene and α-terpeneol. Cumene (C_9H_{12}, b. p. 152 °C) is produced by distilling cuminic acid with lime. Cumaldehyde is used in perfumery. Cumin seed has some anticarcinogenic properties. (Aruna and Sivaramakrishnan, 1992).

Ajowain

The scientific name of ajowain is *Trachyspermum ammi* (Umbelliferae). Three species of ajowain are reported as mentioned below:

(a) *Trachyspermum ammi* (Linn.) Sprague syn
(b) *Trachyspermum copticum* Link;
(c) *Carum copticum* Hiern.

It is known as "ajowan or ajowain" in Bengali and Hindi, as "owa" in Marathi and as "omum" in Tamil. It is called "Bishop's weed" in English.

Description of the Plant

An erect, minutely pubescent, branched annual, stems striated leaves rather distant. 2-3 pinnately divided, segments linear, flower terminal, compound umbels, while small fruit ovoid mucate, aromatic cremocarps (The Wealth of India, 1976).

Uses

Ajowain with its characteristics aromatic taste and pungent smell, is used as a spice in curries. It is used in pickles, certain types of biscuits, confectionary and beverages. The more important use of ajowain is medicinal and its use as a house-hold remedy for indigestion must have led to its use for culinary purposes. Ajowain is much valued for its antispasmodic, stimulant, tonic and carminative properties. It is administered in flatulence, atonicdyspepsia and diarrhoea. It often costitutes an ingredient of cough mixtures. Paste of the crushed fruit is applied for relieving colic pains and a hot and dry fomentation of the fruits on the chest is a common remedy for asthma. Ajowain is used in the preparations of lotions and ointment, applied for checking chronic discharge. The roots of ajowain plant is reported to possess diuretic and carminative properties and are used in febrile conditions and in stomach disorders.

Ajowain also has some antioxidative properties. It has been found that ajowain is a source of natural lipid antioxidant (Mehta, 1994) and antioxidative effect of ajowain in model membrane system has also been reported (Mehta, 1995).

About twenty seven compounds are identified in ajowain of which thymol (61%) is present in highest quantity. Among the other major components p-cymene-15.6% and g-terpinene 11-9% are important (Chialva, 1993)

Ajowain owes it's characteristic odour and taste to the presence of an essential oil (2-4%), ajowain oil. The essential oil obtained by steam distillation of the fruits is known as oil of ajowain (*Ajowain Ka Tel*). It is used in medicine. The oil is the principal source of thymol. Ajowain oil is colourless or brownish yellow liquid possessing a characteristic odour for thymol and burning taste. The principal constituents of the oil are the phenols, mainly thymol (35-60%) and some carvacrol. Thymol easily crystallises from the oil and is known as *Ajowain-Ka-Phool* or *Sat-Ajowain*. The remainder of the oil is called thymene.

Composition of the Ajowain Fruit (Mitra *et al.*, 1969)

Component	*Percentage Composition*
Moisture	7.9 gm
Protein	17.1 gm
Fat	21.8 gm
Fibre	21.2 gm
Carbohydrate	24.6 gm
Minerals	7.9 gm
Composition (mg)/100 gm	
Calcium (Ca)	1.525 mg
Iron (Fe)	27.7 mg
Sodium (Na)	0.56 mg
Potassium (K)	1.390 mg
Thiamine	0.21 mg
Riboflavin	0.28 mg
Nicotinic acid	2.1 mg
Carotene	0.71mg

Thymene which constitutes about 45% of the oil has the following composition and a small amount of champhene and myrcene.

p-Cymene—50-55% used in germicidal purposes.
γ- teprinene—30-35% used in cheap-perfume for soap.
γ-β pinenes—4-5%.
Dipentene—4-6%.

Ajowain oil both pure and dethymolized is employed as an antiseptic and aromatic carminative. It is occasionally used for perfume in disinfectant soap and as an insecticide and also as an expectorant in emphysema, bronchial pneumonia, respiratory ailments and as antidiuretic

Thymol is produced from ajowain oil by treatment with aqueous alkaline solution and regenerating from it by extraction with ether of steam distillation. Thymol is powerful antiseptic and has an agreeable odour. It is useful in variety of fungal infection of the skin. It has been used as an intestinal antiseptic and antihelminthic. It is an ingredient of deodorant, mouthwashes, toothpastes and many pharmaceutical preparations.

Aqueous portion left after the separation of essential oil from steam-distillation is called omum water which is used as carminative. It is believed to be useful in flatulence and gripe, especially in children.

Materials and Methods

Selection of Spices

Cumin seed (*Cuminum cyminum*) and ajowain (*Trachyspermum ammi)* were obtained from the Spices Board, Cochin, India.

Preparation of the Spice Extract

a) Ajowain and cumin were ground separately in mortar pastle.
b) 5 gms of powdered ajowain and cumin were taken in 20 ml alcohol in a closed conical flask.
c) It was kept in a shaker for 2 hours at room temperature, subsequently the same was kept undisturbed overnight.
d) Alcoholic extract was filtered and the filtrate was preserved.
e) The solid residue was re-extracted with alcohol as described above, till the extract recovered as colourless.
f) The pooled filtrate was concentrated by evaporation of solvent and used for further experiments.

Selection of Test Organisms

Antimicrobial activity of ajowain and cumin extract were determined using different micro-organisms.
The test organisms for the present study included

(a) Pathogenic and nonpathogenic bacteria of both plant and animal.
(b) Pathogenic fungi of both plant and animal.
(c) A nonpathogenic yeast.

Testing of Antimicrobial Activities

a) Different volumes of spice extract (ajowain and cumin) were removed from stock solution and placed in different sterile test tubes. The volumes were readjusted with alcohol using sterile pipettes.
b) A fixed volume of 50μl spice extract (equivalent to raw spices, 1–100 mg/ml) was withdrawn from each of the test tubes and subsequently added to the molten nutrient agar medium containing 5 ml of nutrient agar.
c) Agar slopes were made by allowing the agar medium containing spice extract and were inoculated with different test micro-organisms.
d) Control tubes were prepared using equal volume of alcohol instead of spice extract and processed similarly.
e) Agar slopes inoculated with bacteria and yeast were incubated for at least one day at 34°C–37°C in incubator for normal growth.
f) Agar slopes inoculated by fungus were incubated at least 5-6 days at 34°C–37°C in incubator for proper growth.
g) Agar slopes were checked for growth of organisms. The results are expressed as growth of organisms (+) or inhibition of organisms (–) after addition of spice extract or control.

Results and Discussion

Inhibitory Effects of Alcohol Extracts against Bacterial Species

(1) Ajowain [vide Table -1(a)] extract

Bacillus megaterium and *Rhizobium leguminosarum* show positive growth in presence of 1 mg/ml to 18 mg/ml, while no growth is observed from 20 mg/ml. *Sarcina lutea, Micrococcus glutamicum* establish a positive growth upto 16 mg/ml above this concentration show negative growth. *Bacillus cereus* and *Staphylococcus aureus* depict a positive growth upto 12 mg/ml, while growth is inhibited by concentration from 14 mg/ml to 100 mg/ml. *Bacillus licheniformis* shows a positive growth upto 10 mg/ml, while the 12 mg/ml to 100 mg/ml concentration produced growth inhibition. *Escherichia coli, Bradyrhizobium japonicum* and *Bacillus subtilis* have uniform response i.e. positive growth in presence of 1 mg/ml to 8 mg/ml. While the other doses tested i.e. from 10 mg/ml to 100 mg/ml growth is inhibited *Klebsiella aeruginosa* shows negative growth in 8 mg/ml *Agrobacterium tumefaciens* has positive growth in presence of 1.2.4.5. mg/ml of extract while there is a negative growth response when concentrations are 6 mg/ml to 100 mg/ml.

(2) Cumin [vide Table 1(b)] extract

Bacillus cereus establishes differential growth response against cumin extracts. In this case growth is positive in presence of extractive concentration upto 38 mg/ml. *Bacillus megaterium* and *Rhizobium* leguminosarum manifest similar response where growth is negative in presence of extractive concentration of 38 mg/ml to 100 mg/ml. Interestingly, *Bradyrhizobium japonicum* and *Pseudomonas oleovorans* are found to show a similar trend. They grow positively upto 25 mg/ml while no growth has been observed from 26 mg/ml. *Bacillus licheniformis, Agrobacterium tumefaciens* and *Bacillus subtilis*, three bacterial strains show maximum inhibition. First one shows negative growth from 14 mg/ml, while last two show no growth from concentration 10 mg/ml.

Inhibitory Effect against Fungal Species Tested

Growth of all ten fungal species tested in this study was inhibited by ajowain extract at different concentration while six species of fungus, *Aspergillus niger, Sclerotium rolfsii, Fusarium oxysporum,* sensitive *curvularia* sp., *Rhizectonia solani* and *Macrophomina phaseolina* are particularly to the effect of cumin extractive at all concentration (1-100 mg/m!) tested.

(1) Ajowain [vide Table 2 (a)] extract

Aspergillus niger depicts positive growth upto concentration 26 mg/ml and shows negative growth at concentration above 28 mg/ml. Similar response is observed for *Fusarium oxisporum* and *Curvularia* sp. Both shows positive growth upto 20 mg/ml and negative growth when concentration was above 22 mg/ml. Interesting trend is observed for *Rhizectonia solani, Macrophomina phaseolina* and *Trichophyton rubrum.* They establish positive growth only in six test tubes containing concentration 1 mg/ml to 8 mg/ml but depict negative growth from cencentration 10 mg/ml. Effect

TABLE 1 (A) : ANTIMICROBIAL ACTIVITY OF AJOWAIN EXTRACT AGAINST FEW BACTERIAL SPECIES.

Name of Bacterial Species	Ajowain Extract (mg/ml)*														
	0	1	2	4	5	6	8	10	12	14	16	18	20	30	100
Klebsiella pneumonieae	+	+	+	+	+	+	+	+	+	+	+	+	+	+	+
Pseudomonas oleovorans	+	+	+	+	+	+	+	+	+	+	+	+	+	+	+
Pseudomonas aeruginosa	+	+	+	+	+	+	+	+	+	+	+	+	+	+	+
Bacillus megaterium	+	+	+	+	+	+	+	+	+	+	+	+	–	–	–
Rhizobium leguminosarum	+	+	+	+	+	+	+	+	+	+	+	+	–	–	–
Sarcina lutea	+	+	+	+	+	+	+	+	+	+	+	–	–	–	–
Micrococcus glutamicus	+	+	+	+	+	+	+	+	+	+	+	–	–	–	–
Bacillus cereus	+	+	+	+	+	+	+	+	+	–	–	–	–	–	–
Staphylococcus aureus	+	+	+	+	+	+	+	+	+	–	–	–	–	–	–
Bacillus licheniformis	+	+	+	+	+	+	+	+	–	–	–	–	–	–	–
Escherichia coli	+	+	+	+	+	+	+	–	–	–	–	–	–	–	–
Bradyrhizobium japonicum	+	+	+	+	+	+	+	–	–	–	–	–	–	–	–
Bacillus subtilis	+	+	+	+	+	+	+	–	–	–	–	–	–	–	–
Klebsiella aeruginosa	+	+	+	+	+	+	–	–	–	–	–	–	–	–	–
Agrobacterium tumefaciens	+	+	+	+	+	–	–	–	–	–	–	–	–	–	–

(+) = Indicates growth

(–) = Indicates no growth

* The amount of extractive is expressed in terms of dry weight equivalent to the spice from which extraction is made.

TABLE 1 (B): ANTIMICROBIAL ACTIVITY OF CUMIN EXTRACT AGAINST FEW BACTERIAL SPECIES.

Name of Bacterial Species	Cumin Extract (mg/ml)*																								
	0	1	2	4	5	6	8	10	12	14	16	18	20	22	24	25	26	28	30	32	34	36	38	40	100
Staphylococcus aureus	+	+	+	+	+	+	+	+	+	+	+	+	+	+	+	+	+	+	+	+	+	+	+	+	+
Pseudomonas aeruginosa	+	+	+	+	+	+	+	+	+	+	+	+	+	+	+	+	+	+	+	+	+	+	+	+	+
Micrococcus glutamicus	+	+	+	+	+	+	+	+	+	+	+	+	+	+	+	+	+	+	+	+	+	+	+	+	+
Sarcina lutea	+	+	+	+	+	+	+	+	+	+	+	+	+	+	+	+	+	+	+	+	+	+	+	+	+
Klebsiella aeruginosa	+	+	+	+	+	+	+	+	+	+	+	+	+	+	+	+	+	+	+	+	+	+	+	+	+
Klebsiella pneumoniae	+	+	+	+	+	+	+	+	+	+	+	+	+	+	+	+	+	+	+	+	+	+	+	+	+
Escherichia coli	+	+	+	+	+	+	+	+	+	+	+	+	+	+	+	+	+	+	+	+	+	+	+	+	+
Bacillus cereus	+	+	+	+	+	+	+	+	+	+	+	+	+	+	+	+	+	+	+	+	+	+	+	–	–
Bacillus megaterium	+	+	+	+	+	+	+	+	+	+	+	+	+	+	+	+	+	+	+	+	+	+	–	–	–
Rhizobium leguminosarum	+	+	+	+	+	+	+	+	+	+	+	+	+	+	+	+	+	+	+	+	+	+	–	–	–
Bradyrhizobium japonicum	+	+	+	+	+	+	+	+	+	+	+	+	+	+	+	+	–	–	–	–	–	–	–	–	–
Pseudomonas oleovorans	+	+	+	+	+	+	+	+	+	+	+	+	+	+	+	+	–	–	–	–	–	–	–	–	–
Bacillus licheniformis	+	+	+	+	+	+	+	+	+	–	–	–	–	–	–	–	–	–	–	–	–	–	–	–	–
Agrobacterium tumefaciens	+	+	+	+	+	+	+	–	–	–	–	–	–	–	–	–	–	–	–	–	–	–	–	–	–
Bacillus subtilis	+	+	+	+	+	+	+	–	–	–	–	–	–	–	–	–	–	–	–	–	–	–	–	–	–

(+) = Indicates growth
(–) = Indicates no growth
* The amount of extractives expressed in terms of dry weight equivalent to the spice from which extraction is made.

TABLE 2 (A): ANTIMICROBIAL ACTIVITY OF AJOWAIN EXTRACT AGAINST FEW FUNGAL SPECIES.

Name of Fungal Species	Ajowain Extract (mg/ml)*																			
	0	1	2	4	5	6	8	10	12	14	16	18	20	22	24	25	26	28	30	100
Aspergillus niger	+	+	+	+	+	+	+	+	+	+	+	+	+	+	+	+	+	–	–	–
Fusarium oxysporum	+	+	+	+	+	+	+	+	+	+	+	+	+	–	–	–	–	–	–	–
Curvularia sp	+	+	+	+	+	+	+	+	+	+	+	+	+	–	–	–	–	–	–	–
Rhizeconia solani	+	+	+	+	+	+	+	–	–	–	–	–	–	–	–	–	–	–	–	–
Macrophomina phaseolina	+	+	+	+	+	+	+	–	–	–	–	–	–	–	–	–	–	–	–	–
Trichophyton rubrum	+	+	+	+	+	+	–	–	–	–	–	–	–	–	–	–	–	–	–	–
Fusarium sp.	+	+	+	+	+	–	–	–	–	–	–	–	–	–	–	–	–	–	–	–
Alternaria brassicicola	+	+	+	+	+	–	–	–	–	–	–	–	–	–	–	–	–	–	–	–
Helminthosporium oryzae	+	+	+	–	–	–	–	–	–	–	–	–	–	–	–	–	–	–	–	–
Sclerotium rolfsii	+	+	+	–	–	–	–	–	–	–	–	–	–	–	–	–	–	–	–	–

(+) = Indicates growth
(–) = Indicates no growth
*The amount of extractives expressed in terms of dry weight equivalent of the spice from which extraction is made.

on *Fusarium* sp. and *Alternaria brassicicola* are also of similar kind. Both of them grow positive when concentration was below 6 mg/ml above which the extractive caused growth inhibition. *Helminthosporium oryzae* and *Sclerotium rolfsii* growth is inhibited by extractive concentration of 4 mg/ml and above.

(2) Cumin [vide Table 2 (b)] extract

Trichophyton rubrum manifests positive growth upto 25 mg/ml while negative growth takes place when extractive concentration is 26 mg/ml and above. Similar growth response has been found in *Helminthosporium oryzae* and *Fusarium* sp. *Both* show positive growth upto 18 mg/ml, while growth is inhibited from concentration 20 mg/ml to 100 mg/ml. In *Alternaria brassicicola,* negative growth has been observed from concentration 6 mg/ml and above.

Inhibitory Effect against Yeast

Species *Saccharomyces cerevisiae* shows differential growth response for ajowain and cumin extract.

(1) **Ajowain** [vide Table 3 (a)] extract

It shows positive growth upto 8 mg/ml but beyond concentration 10 mg/ml, test tubes are found without any growth.

(2) **Cumin** [vide Table 3 (b)] extract

It shows positive growth in all concentration tested i.e 1 to 100 mg/ml.

Under the condition of experiments, all control tubes contain no ajowain or cumin extract but equivalent amount of solvent (alcohol) (50 μl). At this concentration, the solvent did not produce any growth inhibition of the organisms tested.

Generally speaking, pioneering research during the last three decades have documented many beneficial physiological effects of spices which suggest that spices extend beyond taste and flavour. Digestive stimulant action, hypolipidemic/ hypocholesterolemic efffect, hypoglycaemic potential, prevention and regression of cholesterol gall stones, amelioration of diabetic nephropathy, antioxidant property, antiinflammatory and antiarthritic property, antimutagenic and anticarcinogenic property are major beneficial effects of spices so far documented (Srinivasan, 1998)

Present study includes two spices—ajowain and cumin. Rationale behind their choice has been described earlier and some of their activities are well known. Here, sincere efforts and introspection have been given on their possible antimicrobial activities; and for that 15 bacterial, 10 fungal strains and 1 yeast strain have been selected for study. Alcohol extractives of two spices, ajowain and cumin are tested at different concentrations. Concentration variations have started with 1 mg/ml being lowest to 100 mg/ml being highest.

From the present study, it is clear that *Klebsiella pneumoniae, Pseudomonas oleovorans* and *Pseudomonas aeruginosa* are insensitive against ajowain extract. Rest of the bacteria tested, are sensitive to this extractive. The clubbing of *Bacillus megaterium* and *Rhizobium leguminosarum, Sarcina lutea* and *Micrococcus glutamicus. Bacillus cereus* and *Staphylococcus aureus; Escherichia coli,*

TABLE 2 (B): ANTIMICROBIAL ACTIVITY OF CUMIN EXTRACT AGAINST FEW FUNGAL SPECIES.

Name of Fungal Species	Cumin Extract (mg/ml)*																			
	0	1	2	4	5	6	8	10	12	14	16	18	20	22	24	25	26	28	30	100
Aspergillus niger	+	+	+	+	+	+	+	+	+	+	+	+	+	+	+	+	+	+	+	+
Sclerotium rolfsii	+	+	+	+	+	+	+	+	+	+	+	+	+	+	+	+	+	+	+	+
Fusarium oxysporum	+	+	+	+	+	+	+	+	+	+	+	+	+	+	+	+	+	+	+	+
Curvularia sp.	+	+	+	+	+	+	+	+	+	+	+	+	+	+	+	+	+	+	+	+
Macrophomina phaseolina	+	+	+	+	+	+	+	+	+	+	+	+	+	+	+	+	+	+	+	+
Rhizectonia solani	+	+	+	+	+	+	+	+	+	+	+	+	+	+	+	+	+	+	+	+
Trichophyton rubrum	+	+	+	+	+	+	+	+	+	+	+	+	+	+	+	+	–	–	–	–
Fusarium sp.	+	+	+	+	+	+	+	+	+	+	+	+	–	–	–	–	–	–	–	–
Helminthosporium oryzae	+	+	+	+	+	+	+	+	+	+	+	+	–	–	–	–	–	–	–	–
Alternaria brassicicola	+	+	+	+	+	–	–	–	–	–	–	–	–	–	–	–	–	–	–	–

(+) = Indicates growth
(–) = Indicates no growth
*The amount of extractive is expressed in terms of dry weight equivalent of the spice from which extraction is made.

TABLE 3 (A): ANTIMICROBIAL ACTIVITY OF AJOWAIN EXTRACT AGAINST YEAST SPECIES.

Name of Yeast Species	*Ajowain Extract (mg/ml)**									
	0	1	2	4	5	6	8	10	12	100
Saccaharomyces cerevisiae	+	+	+	+	+	+	+	–	–	–

(+) = Indicates growth
(–) = Indicates no growth
* The amount of extractive is expressed in terms of dry weight equivalent to the spice from which extraction is made.

TABLE 3 (B): ANTIMICROBIAL ACTIVITY OF CUMIN EXTRACT AGAINST YEAST SPECIES.

Name of Yeast Species	*Cumin Extract (mg/ml)**									
	0	1	2	4	5	6	8	10	12	100
Saccaharomyces cerevisiae	+	+	+	+	+	+	+	+	+	+

(+) = Indicates growth
(–) = Indicates no growth
* The amount of extractive is expressed in terms of dry weight equivalent to the spice from which extraction is made.

Bradyrhizobium japonicum and *Bacillus subtilis* are quite interesting as far as their Minimum Inhibitory Concentrations are concerned. On the other hand, *Bacillus licheniformis, Agrobacterium tumefaciens* and *Klebsiella aeruginosa* have comparable sensitivity towards this extract. These results are supported by previous reports which includes those like antibacterial root canal filling materials effect of thymol (Seven, 1989, Geockariz, 1974, Meeker, 1988 Ross, 1985) terpens and terpene derivative has been shown to possess antimicrobial and verminoxious property. Thymol in modern mouthwash is incorporated for its assumed antimicrobial activity and antimicrobial effects on endodontic filling cements on microorganisms from root canal by ajowain (Pupo, 1983).

The growth of bacterial species namely *Staphylococus aureus, Pseudomonas aeruginosa, Micrococcus glutamicus, Sarcina lutea, Klebsiella pneumoniae* and *Escherichia coli* are all insensitive to all the concentration of cumin tested. The present study shows that the growth of *Bacillus megaterium* and *Rhizobium leguminosarum, Bradyrhizobium japonicum, Pseudomonas oleovorans, Agrobacterium tumefaciens* and *Bacillus subtilis* are effecting at similar level of Minimum Inhibitory Concentration. *Bacillus cereus, Bacillus licheniformis* are also exclusive individually so far as their expression of Minimum Inhibitory Concentrations are concerned. These results are also well recommended by the work of previous scientists which includes those like cumin inhibits the growth and acid production of *Lactobacillus plantarum* and *Leuconostic mesenteroids* (Kivanc, 1991) and essential oil of cumin shows various degrees of growth inhibition for *Mycoderma, Lactobacillus, Bacillus cereus* (Meena and Sethi, 1994) etc.

As mentioned earlier, bacterial strains that are used in the present study are either pathogenic and nonpathogenic but one has to take special interest about pathogenic bacteria which could cause diseases and other harms. For example,

Klebsiella pneumoniae can cause pneumonia, respiratory tract infection etc. *Pseudomonas aeruginosa* causes meningitis, nosocomial infection etc. *Sarcina lutea* causes mammalian intestinal tract infection; *Bacillus cereus* causes food poisoning. *Staphylococcus aureus* causes boils, pustules, pimples, furucles, abscesses, carbuncles, *Escherichia coli* causes diarrhoeal disease, bacterimia, *Bacillus subtilis* causes eye lesions; *Agrobacterium tumefaciens*, a plant tumor causing pathogen.

Above bacterial strains clearly show their potentiality as pathogen. In this context, the present study gains a further dimension as the control of growth of these pathogen by ajowain and cumin extract, that are used in this study obviously depict their potentiality as control agents of these bacteria. Moreover, it won't be out of place to mention here that for the first time these bacterial species are tested against the above mentioned spices.

The present study establishes the phenomenon that ajowain cumin can well be recommended against those strains where they show sensitivity from certain level of concentration. For exmple, ajowain is particularly effective against *Bacillus megaterium* to *Agrobacterium tumefaciens.*

Cumin is particularly effective in its action against specially *Bacillus cereus* and *Bacillus subtilis.* An interesting observation emerges comparing the activity of ajowain and cumin against bacteria. Both the spices are growth inhibitory to species like *Bacillus cereus, Bacillus megaterium. Rhizobium leguminosarum, Bradyrhizobium Japonicum, Bacillus licheniformis, Agrobacterium tumefaciens* and *Bacillus subtilis* though their Minimum Inhibitory Concentration vary.

Interestingly, from the above mentioned experimental findings, it has been quite clear that all the fungal strains tested show sensitiveness against different concentration of ajowain tested, though their Minimum Inhibitory Concentration vary appreciably.

Fusarium oxysporum and *Curvularia* sp., *Rhizectonia solani, Macrophomina phaseolina* and *Trichophyton rubrum, Fusarium* spp. and *Alternaria brassiciola*, *Helminthosporium oryzae* and *Sclerotium rolfsii* show similar type of inhibition pattern. These results are well supported by some reports like, ajowain shows inhibitory effect on some food borne fungi (Meena and Sethi, 1994). Inhibitory effect of ajowain on growth and toxin production by toxigenic fungi is reported. It shows complete inhibition of *Aspergillus flavus* and *Aspergillus versicolor* (HitoKoto, 1980) growth. Reports are also available on the inhibition in aflatoxin formation by ajowain (Mabrouk, 1980) and fungitoxic properties of the essential oil of ajowain (Singh, 1986).

On the other hand, *Aspergillus niger, Sclerotium rolfsii, Fusarium oxysporum, Curvularia* sp., *Rhizectonia solani* and *Macrophomina phaseolina* are insensitive to different concentrations of cumin extracts.

Helminthosporium oryzae and *Fusarium* spp. show sensitiveness and have similar kind of Minimum Inhibitory Concentration and *Trichophyton rubrum* and *Alternaria brassicicola* are exclusive in relation to Minimum Inhibitory Concentration.

These results support the previous reports that inhibition of aflatoxin formation of cumin (Mabrouk, 1980), inhibitory activity towards the growth, respiration, lipase and mycotoxin production by *Aspergillus paraciticus* and *A. fumigatus* (Akgul, 1988).

Cumin inhibits the growth of some food spoilage fungi (Thyagaraja *et al.*, 1996). Inhibitory effect of essential oil of cumin on growth of food containing fungi (Pauli,

1987). Cumin controls *Alternaria blight* and powdery mildew disease, (Geemawat, 1969). Cumin oil is helpful for disorders of skin functioning, e.g. inflammation, acni, eczema, scabies, blunt trauma, skin mycosis, haematoma, nermodermititis and psorias. It covers different types of allergic contact dermatitis (Steinmann, 1997).

Alternaria brassicicola causes leafspot on crucifers, *Aspergillus niger* if inhaled causes allergy and even lung infection. *Curvularia* sp. causes leafspot on grasses; cereals, *Helminthosporium oryzae* causes leaf brown spot of paddy plant. *Macrophomina phaseolina* causes root and stem rot of jute etc. *Fusarium* sp. causes seedling disease of cereals, root rot, wilting of several plants. *Sclerotium rolfsii* causes damping of seedling, stem canker, crownblight of legumes, crucifers. *Rhizectonia solani* causes scurf disease of potato, *Trichophyton rubrum* causes athelet's foot disease in humans.

The above mentioned growth inhibitory activities of ajowain and cumin extracts against the fungal strains have clearly established their potential usage for the purpose of controlling fungal infection and it could be very productive, particularly when such strains are very much harmful in everyday life.

Yeast, *Saccharomyces cerevisiae* being nonpathogenic is only taken for the purpose of academic interest as already mentioned. Interestingly, ajowain and cumin show different effects in their level of activities. Essential oil of ajowain shows various degrees of inhibition against *Saccharomyces cerevisiae*. (Neena and Sethi, 1994)

Finally, it can be concluded that both the spices selected for the study have been proven to be exceptionally good and effective against the different damaging microbial agents. This gains further significance when both the spices are found in plenty in our country which being tropical, experiences several diseases and other damaging problems from different microbial agents. So, further, well directed studies of this nature are highly recommended to establish ajowain and cumin as therapeutic agents for the treatment of microbial diseases.

References

Akgul, A. and Kivanc, M., 1988. Inhibitory effects of selected Turkish spices and organo components on some food born fungi. *Int. J.Food. Microbiol.* (Netherland), **6** (3): 263.

Aruna, K. and Sivaramakrishnan, V. M., 1992. Anticarcinogenic effect of some plant products. *Food. Chem. Toxicol.* **30** (III): 935.

Chialva, F., 1993. Essential oil constituents of *Trachyspermum copticum,* (L) Link fruits, *J. Essent. Oil. Res. Carol Steam III.* Allured Publishing Corporation **5** (I): 105.

Davis, B. D., 1980. *Microbiology* (3rd edn). Harper and Row Publisher. pp. 1355.

El-Fatatny, H. M., 1975. Isolation and structural assignment of an antimicrobial principle from the volatile oil of *Nigella sativa* L. Seed. *Pharmazie.* (GDR), **30** (2): 109.

Gemawat, P. D., 1969. Control of *Alternaria* blight and powdery mildew by cumin. *Hindustan Antibiel. Bull.* (India) **II** (3): 169.

Gockaritz, D., 1974. Terpenes and tarpene derivatives of carvone and champhor type their antimicrobial and verminoxious properties, *Pharmazie* (GDR). **29** (5): 339.

Hitokoto, H., 1980. Inhibitory effects of spices on growth and toxin production of toxigenic fungi. *Appl. Environ. Microbiol,* (USA) **39** (4): 818.

Kivanc, M., 1991. Inhibitory and stimulatory effects of cumin oregano and their essential oil on growth and acid production of *Lactobacillus plantarum* and *Leuconostoc mesenteroides. Int. J. Food Microbiol.* (Netherland) **13** (1): 89.

Mabrouk, S. S., 1980. Inhibition of aflatoxin formation by some spices. *Z. lebensum. Unters, Forsch.* (FDR) **171** (5): 334.

Meeker, H. G., 1988. The antibacterial action of Eugenol thymeoil and related essential oil used in dentistry. *Compendium (USA)* **9** (1): 32, 34, 38.

Meena, M. R. and Sethi, Vijay, 1984. Antimicrobial activity of essential oil from spices. *J. Food. Sci. Technol.*, **31** (I): 68.

Mehta, R. L., 1994. Ajowain as source of natural lipid antioxidant. *J. Agric. Food. Chem.* Washington D. C. American Chemical Society, **42** (7): 1420.

Mehta, R. L., 1995. Antioxidative effect of ajowain in a model system. *J. Am. Dit. Chem. Soc.*, Champaign IL ADCS Press **72** (10): 1215.

Mitra, S. N., Roy, B. and Bose, P. K., 1969. A note on the composition of ajowain (Bishop's Weed). *J. Inst. Chem.* **41**: 108.

Pauli, A., 1987. Inhibitory effects of essential oil components on growth of food containing fungi, *Z. lebensm. Ulters, Forsch. (*FDR*)*, **185** (I): 10.

Pleczar, M. J., Chan, E. C. S. and Noel, R., 1986. *Krieg Microbiology* (5th ed.). Tata Mc Graw-Hill Publishing Company Limited, New Delhi: 118.

Pruthi, J. S., 1979. Spices and Condiment (2nd ed.). National Book Trust India. : 103.

Pupo, J., 1983. Antimicrobial effects of endodontic filling cements on microorganisms from root canal. *Oral. Surg. Oral. Med. Oral. Pathol* (USA), **55** (6): 622.

Ross, N. M., 1985. The antimicrobial activity of modern mouthwashes. *British, Dent. J.* (England), **158** (7): 241.

Seven, N., 1989. The characteristics of antibacterial root canalfilling materials, *Mikrobiyol. Bul.* (Turkey) **23** (4): 361.

Singh, S. P., 1986. Fungitoxic properties of the essential oil of *Trachyspermum ammi* sprague. *Mykosen* (FDR) **29** (I): 37.

Srinivasan, K., 1988. Spices valued for more than taste and flavour of foods. *Proceedings National Conference of Recent Trends in Spices and Medicinal Plants Research* and *11th Annual Meeting of the Acad. of Plant Sci. India* B 20.

Steinmann, Schatzle, A. M., Agatha, M., Breet, R., 1997. Allergic contact dermatitis from black cumin after tropical use, *Contact Dermatitis* **36** (5): 268.

Thyagaraja, N. and Hosono, A., 1996. Effect of spice extract on fungal inhibition, *Lebensmittel, Wissenschaft. Technologie.* **29** (3): 286.

1950. *The Wealth of India, A Dictionary of Indian Raw materials and Industrial Products.* CSIR, New Delhi Vol II (C): 427.

1976. *The Wealth of India, A Dictionary of Indian Raw Materials and Industrial Products.* CSIR, New Delhi, Vol X-SP-W: 267.

Recent Trends in Spices and Medicinal Plants Research pp. 62-65, 2000
Ed. Amit Krishna De
Associated Publishing Co., New Delhi

1.7

SPICES AND MEDICINAL PLANTS IN TRIBAL MEDICINES OF BANKURA DISTRICT, WEST BENGAL

Ramsankar Basu[*] and Prasanta Kumar Mukherjee[**]

Abstract

The district of Bankura has an approximate resemblance of an isosceles triangle in West Bengal.This area is lying under 22°38'—23°38'N latitude, 86°36'—87°46' E longitude, acquires a rural area of 6,814.3 sq.km., and town area of 66.7 sq.km. Total population in the district is approximately 25,00,000. Rural population is 92.53%; of which tribals are 12.58%. Tribals are Santhals, Bhumijs, Mudikoras, Mundas, Kherias and Bedias. Forest covers an area of 3,33,00 acres or 21.54% of the total area. The district forms a part of the eastern fringe of Chhota Nagpur plateau with isolated spurs, ridges and rolling uplands to an elevation of 445.92 m in the North and West. The climax vegetation is " Northern Tropical Dry Deciduous" type. The jungles produce a quantity of medicinal plants which are being used by the natives including the tribals as their medicament. Some secluded jungle areas of this district have been surveyed from 1995 by the authors in search of medicinal plants. 146 such plant species alongwith 13 contributing as spices are dealt with in this paper.

Introduction

The district of Bankura in West Bengal lies under the latitude 22°38'—23°38' N and longitude 86°36'— 87°46' E, acquires an area of 6,881 sq.km., make an intermediate tract in between the rice producing alluvial plains of Bengal to the East and Chhota Nagpur plateau on the West. The climate is sultry and hot in summer, having rains due to the south-west monsoon during June and September. Annual rainfall ranges between 1,101 and 1,400 mm. The highest maximum temperature is 47.2°C in May and the minimum temperature is 5°C in February. North-Westerlies come from the north–west direction. Forest covers 1,404 sq.km in all, which works out to 21.54% of the total area of this district. The climax of vegetation is "Northern Tropical Dry Deciduous" types in India (O'Malley, 1908).

Tribal population in this backward district area are 12.58%. Tribals are 'Santhals' and 'Bhumijs' of Austro–Asiatic Kolarian chain, 'Mudikoras' and 'Mundas' of Austral class, 'Kherias' of Dravidian and 'Bedias' of Aryan race. They prefer to live in

[*]Botany Department, Achruram Memorial College, Jhalda - 723 202, Dt.Purulia, West Bengal
[**]Botany Department, University of Calcutta, Calcutta-700019

remote hilly jungle areas and depend upon forest resources in one form or others (Anonymous, 1995; Dalton, 1872). The jungles produce a quantity of medicinal plants which are used by the natives as their medicament along with some spices. They also sell up these mass of collections in return for prices or something paid in exchange. Due to lack of actual outlets for marketing, the tribals fall prey to deceits by the middlemen.

Discussion

The uses of different plants, their preparations and dosages are noted below:

Abortion in Female

In 5 gms of tuber of *Gloriosa superba* (*Languli late*) are added only 2 gms of gum resins of the roots of *Ferula jaeschkeana* (*Multani Hing*), 25 gms roots of *Moringa oleifera* (*Sajina*), 5 gms roots of *Achyranthes aspera* (*Apang*), boiled in 500 ml of water and volume reduced down to 250 ml; and sieved. A cup amount of the mixture is given orally once daily for three days. It induces the expulsion of foetus even of 8-9 months old.

Asthma of Cattle

Plants of *Andrographis paniculata* (*Kalmegh*), roots of *Aristolochia indica* (*Ishwarmul*), bark of *Erycibe paniculeta* (*Kari*) either in fresh or in dry condition mixed equally; and to it are added a little bit of the stem of *Glycerrhiza glabra* (*Jastimadhu*), 3 seeds of *Piper nigrum* (*Golmorich*) seeds of *Trachyspermum ammi* (*Jowan*) and digestive salt (*beetnoon*); mixed thoroughly. If administered orally once a day on alternate days for a month; the asthmatical disorders may be cured.

Blood Dysentry (Antrik) of Men and Live-stock

To a little amount of the seeds of *Nigella sativa* (*Kalajira*) the following are added: seeds of *Trachysperum ammi* (*Jowan*), *Foeniculum vulgare* (*Mouri*), *Trigonella faenum-graecum* (*Methi*), *Amomum subulatum* (*Boraelach*), flower buds of *Syzygium aromaticum* (*Long*), resin of *Shorea robusta* (*sal*) alongwith the roots of *Hymenodictyon orixense* (*Jhurkul*), *Atylosia scarabaeoides* (*Ben Kurthi*), *Achyranthes aspera* (*Apang*), bark of *Lannea coromandelica* (*Doka*), *Syzygium cumini* (*Jam*), *Terminalia tomentosa* (*Asta*) and candy sugar. All are pressed, grounded and mixed. Fed orally 3 times a day for three days to cure blood dysentry.

Chronic Typhoid Fever (Kala-Sannipati)

Whole plant of *Hedyotis corymbosa* (*Khet-papra*) pasted with some seeds of *Trachysperum ammi* (*Jowan*) and digestive salt (*beet-noon*). A cup amount of the paste given orally at 4 to 6 hours interval for 3 days for rapid recovery.

Foot and Mouth Disease of Cattle, even for Men

Bark of *Holarrhena pubescens* (Kurchi), seeds of *Nigella sativa* (*Kalajira*), bulb of *Allium sativum* (*Lashun*) shreded, grounded and mixed in mustard oil (from the seeds of *Brassica nigra)*. The mixture is warmed gently and applied on the sores for recovery.

Indigestion, Acidity, Liver-disease, Stomachic Ulcers

Bark of *Butea superba* (*Lat palash*), *Buchanamia lanzan* (*Pial*), *Terminalia chebula* (*Haritaki*), *Terminalia bellirica* (*Bahera*), *Phyllanthus emblica* (*Aonla*), rhizome of *Zingiber officinale* (*Ada*), seeds of *Trachyspermum ammi* (*Jowan*), *Vernonia anthelmintica* (*Somraj*), *Hollarhena pubescens* (*Kurchi*) are crushed together to make a tonic and administered orally for curing these chronic diseases.

Leucorrhoea of Female and Spermetorrhoea of Male

Exuded juice of the stems and flowers of *Butea monosperma* var. *lutea* (Yellow flowered Palas) and *Eriodendron anfractuosum* (yellow flowered *simal*) are mixed proportionately; roots of *Smilex Zeylanica* (*Ramdantan*), *Asparagus racemosus* (*Satamul*), *Curculigo orchioides* (*Talamuli*), *Hemidesmus indicus* (*Anantamul*), rhizomes of *Nelumbo nucifera* (*Padma*), *Costus speciosus* (*Kiya-kanda*), tuber of *Pueraria tuberosa* (*Patal dingla*),stem of *Tinospora cordifolia* (*Gulancha*), seeds of *Seseli indicum* (*Ban-jowan*), *Cuminum cyminum* (*Jira*), *Piper nigrum* (*Golmorich*) are then mixed together and after addition of candy, sugar tablets are made. One tablet/day given orally in empty stomach for 11 days. In case of male, instead of *Nelumbo*, roots of *Buettneria herbacea* (*Kamraj*), *Hygrophila auriculata* (*Talmakhna*) are pounded and added.

Loose-evacuation of Cattle

A mixture is made with the following ingredients: Roots of *Asparagus racemosus* (*Satamul*), rhizome of *Curcuma longa* (*Haldi*) and *Nelumbo nucifera* (*Lal-shaluk*), fruits of *Diospyros malabarica* (*Makar-kend*), seeds of *Trachyspermum ammi* (*Jowan*), leaves of *Anthocephalous chinensis* (*Kadam*) and *Bambusa arundinacea* (*Bans*) and digestive salt (*beet noon*). The mix is added in fodder and fed intermittently to the cattles.

Stomachic Pain

Fruits of *Helicteris isora* (*Ga-machra*), *Ventilago denticulata* (*Bongasarjom*) and seeds of *Nigella sativa* (*Kalajira*) are fried in mustard oil (from the seeds of *Brassica nigra)*. The oily mixture is massaged over the belly for relieving pain.

Conclusion

Inhabitants of the jungle areas, having life below the poverty line, collect the herbs, roots and rhizomes, barks, fruits and seeds etc. Local sell out value per 250 gms of herbage are Rs. 2.50. Earned value per individual per annum are Rs. 1000-3000. This earned value may rise upto Rs. 2000-4000 if they get the chance to sell out the herbs alongwith the spices directly by organising themselves in co-operatives.

And, with that, if they are trained for cultivation of these spices plants, further improvements to their economy can be brought about.

To get the cost of benefit, in favour of tribal people, the following steps are suggested.

(i) Medicinal plants which have more monetary value and which occur naturally in their areas should be selected and their method of cultivation and exploitation should be standardized.

(ii) Forest Protection Committees and the Panchyats should take interest in this respect.

(iii) Awareness among the tribal/rural people for the conservation, development and recycling of forest resources is necessary and NGOs should pay their attention to this aspect.

Acknowledgement

We are thankful to the authorities of DST, Govt. of India for providing financial assistance. We are grateful to the Divisional Forest Office (N & S), Bankura for providing facilities in the field study.

References

Anonymous, 1995. Bankura District Statistics. B. G. Press, Govt. of West Bengal, Calcutta -700 027.

Bennet, S. S. R., 1987. Name Changes in Flowering Plants of India and Adjacent Regions. Triseas Pub. Dehra Dun -248 001.

Dalton, E. T., 1872. Descriptive Ethnology of Bengal; Indian Studies; Past & Present, Calcutta-700 020.

O' Malley, L. S. S., 1908. Bengal District Gazetteers, Bankura. The Bengal Secretariate Book Dept. Calcutta.

Recent Trends in Spices and Medicinal Plants Research pp. 66-70, 2000
Ed. Amit Krishna De
Associated Publishing Co., New Delhi

1.8

ETHNOMEDICINAL PLANTS USED AS ANTIDOTE FOR POISONOUS BITES AMONG THE TRIBALS OF SOUTHERN WESTERN GHATS, INDIA

D. Thangadurai*

Abstract

The paper deals with the plants used as antidote against poisonous bites practiced by the tribals of Kalakad Mundanthurai Tiger Reserve, in Tirunelveli district of Tamilnadu in Southern Western Ghats. Plants used for curing poisonous bites such as insect-bite, leach -bite, rat-bite, scorpion sting and snake–bite, have been well documented here.

Introduction

Ethnobotany deals with the study about tribal and rural people for recording their unique knowledge about plant wealth and for search of new sources of herbal drugs, edible plants and other related aspects (The Wealth of India). Plants have been used as a source of medicine throughout the world for more than 5000 years and still continue to occupy an important place in traditional as well as modern system of medicine. About 2000 plants species have been found to possess the medicinal value in all the four system of indigenous medicine viz. Ayurveda, Unani, Siddha and Homeopathy. This has a silver lining for a country like ours which possesses a vast resource of medicinal plants (Kumar & Mathur, 1992).

Materials and Methods

Kalakad Mundanthurai Tiger Reserve, is 17th in the twenty two Tiger reserves declared in India. The Reserve lies in between latitudes of 8° 25" and 8° 53" North and longitudes of 77° 10° and 77° 85" East in Tirunelveli district. The region lies in one of the 18 *Hot Spots of the World* (Myers, 1990) and one of the 5 Centres of Plants Diversity in India (WCMC, 1992). The region is rich in floristic wealth, with 2105 species in 872 genera belonging to 157 families and medicinal wealth. In this

*Sir Paramakalyani Centre for Environmental Sciences, Mononmaniam Sundaranar University, Alwarkurichi 627 412, Tamilnadu.

reserve, a jungle tribe known as *Kanis* or *Kanikars*, who are believed to have been migrated from Kerala to this region to do plantation work, timber extraction, honey collection and hunting during the British Period (Thurston and Rangachari, 1909) are living in Servalar, Agastyar Kanikudiyuruppu, Chinnamylar, Periyamylar and Ingikulli, with a total population of about 499 members in 117 families. The tribals mostly depend on the forests for their survival as well as for curing their ailments. *Periyamylar* and *Ingikulli Kanis* are exclusively dependent on the plants for all their activities even now.

Modern medicine has no satisfactory remedies for poisonous bites. It has been found that numerous plants are used as antidote against poisonous bites by these tribals as medicine. They use various parts of the plants such as root, stem, bark, gum leaves, fruits seeds, flowers etc.

Ethnomedicinal data collection were made in different tribal areas of Kalakad Mundanthurai Tiger Reserve of Tirunelveli district in Tamilnadu. During that time, some interesting tribal uses of plants as antidote against poisonous bites have come to light. Information regarding the uses of these species were obtained by personal interview with tribal people, forest dwellers and other local inhabitants. The traditional uses of various plants for various poisonous bites have been recorded here. The record information were compared with the uses reported in the earlier literatures (Chopra *et al.*, 1956 and 1969; Dutta, 1877; Gamble, 1915; Jain, 1981; Kirtikar and Basu, 1935).

Results

The following list provides botanical name of the plant and its family name followed by vernacular name(s) used by the tribals of the area and mode of use.

Acalypha indica Linn.

Euphorbiaceae
Kuppaimeni

Plant leaves are mixed with *chunnam* (calcium carbonate) grounded into paste and applied on snake bitten wounds as an antidote for poison. If wound is invisible, it is administered internally.

Androqraphis paniculata Ness.

Acanthaceae
Nilavembu/sirianangai

A tea spoonful of either fresh or dried leaf powder is taken internally with water both in the morning and evening as an antidote for all poisons.

Ageratum conyzoides Linn.

Asteraceae
Pum-Pillu

Fresh leaves are grounded and the juice is applied to leach bite portion to stop bleeding, also used for cut.

Aristolochia tagala chan. Syn. *A. roxburghiana* Klotz.

Aristolochiaceae
Malaiarasan

Whole plant extract is used as antidote for poison.

Aristolochia bracteolata Lamk Syn. *A. bracteata* Retz.

Aristolochiaceae	Leaves and roots are used as antidote for poison. Plants are used as insecticide or insect repellent due to the presence of aristolochic acid.

Aristolochia indica Linn.

Aristolochiaceae *Perumarundukodi*	Grounded leaves and roots are mixed in a glass of water taken orally to cure snake-bite and scorpion sting.

Boerhaavia diffusa Linn.

Nyctaginaceae *Puranavathai*	The roots are crushed and applied over the affected part in snake bite for relief.

Calotropis procera (Ait.) R. Br.

Asclepiadaceae *Vellerku*	Latex is applied over the affected parts as an antidote to snake bite.

Cassia occidentalis Linn.

Caesaloiniaceae *Ponnavirai*	The root is useful in curing scorpion sting and snake-bite.

Citrullus colocynthis Schard

Cucurbitaceae *Paycumuti*	Paste made out of the mixture of fruit and root is used as antidote to snake poisons.

Cocculus cordifolius Miers

Menispermaceae *Sindikodi*	15-20 gms of the root is a good emetic and is used as a successful remedy in snake-bite.

Cycas circinalis Linn.

Cycadaceae *Salapanaikai*	Rhizome is used as antidote for poison.

Entada scandens Benth.

Leguminoceae *Yanainerungi/Thallikkai*	Seed endosperm is used as antidote for poison.

Gloriosa superba Linn.

Liliaceae *Kazhappaikizhangu*	Rhizome is immersed in salt and buttermilk continuously for seven days and taken out and dried under sunlight. Five gms of the powder is administered for snake-bite, scorpion-sting and millepid poison.

Indigofera longeracemosa Boiv.Ex Baill.

Fabaceae *Karunelli*	Root paste is applied on the place where the snake has bitten.

Piper betle Linn.

Piperaceae *Vettrilai*	Juices of leaves alongwith milagu (*Piper nigrum*) is taken internally for twenty one days as an antidote for millepid poison.

Polyqala elongata Klein.

Polygalaceae *Periyanangai*	Whole plants extract is used as antidote for poison.

Rauwolfia serpentina Benth.

Apocynaceae *Chivanmelpodi*	Root is used as an antidote for snake-bite and insect-bite.

Ricinus communis Linn.

Euphorbiaceae *Amanakku*	Seeds are used in curing scorpion sting.

Sapindus emarginatus

Sapindaceae *Poongankottai*	Leaf paste is used as antidote for poison.

Swertia chirata Ham.

Gentianaceae *Nila-vembu*	Whole plant extract is used as antidote against snake-bite.

Tecoma stans (Linn.) Kunch

Bignoniaceae *Sonnapatti*	Root paste is used to cure snake-bite, rat-bite and scorpion sting.

Trichodesma zeylanicum R. Br.

Boraginaceae *Kavuthumbai*	It has the ability to neutralize the effect of sixty four types of poisons.

Vernonia cinerea Less.

Compositae *Naichette*	Leaf paste is given in curing insect-bite.

Wrightia tinctoria R.Br.

Apocynaceae *Veppalai*	Bark paste is an antidote to snake poison for goats.

Discussion

The medicinal uses of few species reported in the this area have not yet been mentioned in any of the published literatures and pharmacopoeias. The tribal uses of above discussed plants are potential source of medicine, leaves the scope for further research and needs further assessment in laboratory before undertaking any large scale cultivation and consumption. The knowledge of the society can be used for welfare of the larger society of the world in general and India in particular.

References

Chopra, R. N., Nayar, S. L. and Chopra, I. C.,1956. *Glossary of Indian Medicinal Plants,* CSIR, New Delhi.

Chopra, I. C. and Varma, B. S., 1969. *Supplement to Glossary of Indian Medicinal Plants,* New Delhi.

Dutta, V. C., 1877. *The Materia Medica of the Hindus,* Calcutta.

Gamble, J. J., 1915. *Flora of the Presidency of Madras,* Calcutta.

Jain, S. K., 1981. *Glimpses of Indian Ethnobotany,* Oxford & IBH Publishing Co. New Delhi.

Kirtikar, K. R. and Basu, B. D., 1935. *Indian Medicinal Plants,* 2nd edition. Allahabad.

Kumar, S. and Mathur, R., 1922. Perspectives in tribal medicines with special reference to some *Asteraceous* plants in India. *J. Econ. Tax. Bot.* Addl. Ser. **10:**19-22.

Myers, N., 1990. The biodiversity challenge: expanded hot spots analysis. *The Environmentalist,* **10:** 243-256.

Thurston, E. and Rangachari, K., 1909. *Castes and Tribes of South India.* Vol IV: 162-177. Government Press, Madras (Repr.ed. 1975, Cosmo Publications, Delhi).

The Wealth of India, 1948-1976. *A Dictionary of Indian Raw Material and Industrial Products. Raw Material Series.* CSIR, New Delhi.

World Conservation Monitoring Cente, 1992. *Global Biodiversity Status of the Earth's Living Resources.* pp xx+594. Chapman & Hall, London.

Chapter 2

Cultivation: Micropopagation: Tissue Culture

Recent Trends in Spices and Medicinal Plants Research pp. 73-77, 2000
Ed. Amit Krishna De
Associated Publishing Co., New Delhi

2.1

EXPLOITING FOREST AND PLANTATION PRODUCTS

B. R. Roy*

Abstract

Many types of activities particularly prospecting for chemicals, biotechnology and patents, etc. these days attach to exploitation of plants. Similar activities can also be applied to marine organisms. A few of them have been discussed in this chapter.

Spices—The Agenda for Action

Spices had been a more important item of commerce of India in ancient days than these days. India's share of international trade of spices declined for our relative inaction on the one hand and other countries' advancement on the other. It is a nice field and India must think of the following:

Production

There is a very large number of spices available in the world trade. But only a small number is in vogue in India. This number is also much less than that of the whole list the Spice Board has been assigned to deal with. India having all kinds of climate, altitude, soil and geographical regions can grow almost everything, especially with the help of agricultural genetics. But it stuck to what had been made available by tradition which means that no person or group consciously did any promotion work and nevertheless some tangible benefit accrued over long years, just as the tribals casually gathered information we now grab to benefit from. Hence, there should be a conscious developmental effort to produce all the spices for internal consumption and for export. Productivity should be the maximum possible; why should we languish below what others account for ? Methods for new crops and higher productivity are known; we are to apply ourselves faithfully to the task of improving.

India exports only 5% of its production of spices. While this is not a mean activity, production should be much more, given India's advantages. Not only all known spices should be produced and handled, new ones be researched and adopted

*Ex-Director, Central Forensic Laboratory, EC-184, Salt Lake, Calcutta - 700 064

from the multitude of ethnic uses particularly of the tribals. The latter's tradition is comparable to what a huge investigational input can do presently.

Quality Improvement

Though not inalienably connected with nutrition and food processing, spices are favourites in cooking and household food preparation. Agricultural marketing included some spices for Agmark and food regulation in Prevention of Food Adulteration Rules recognised some common spices. The specifications demanded standard qualities; these gave fillip to quality assurance. However, quality improvement should be a continuous pursuit through beneficial agronomic practices, improved strains and right post-harvest technology. General and special quality improvement programmes should be undertaken for a product of competitive potential at the world markets.

Technolgy

The whole spices and particulary the products thereof suffer in quality for infestation by fungus and insects and for deficiency of essential oils. This happens for inadequate post-harvest technology and less than ideal storage and transport. Controlled atmosphere storage, modified atmosphere storage and storage under nitrogen or carbondioxide specially for oxidisable products like oleoresins can be recommended. Getting pulverised products under cryogen not only prevents oxidation and volatilisation of essential oils but also the hazard of explosion.

Value-added products from spices are specially appropriate for export. Manufacture of oleoresins of top quality should be looked into extensively for larger number of spices and intensively for better organoleptic and chemical qualities. Extracts not much used in India are an accepted commodity internationally; this can also be produced specially for exports.

Spice powders, popular as they are in India, suffer from loss and modification of flavour. Retention of the fresh flavour and prevention of oxidation of the useful principles, hygroscopicity and infestation must be ensured for good quality products. Flavours and extracts can be tried for enrichment of restoration of the derived products. Spice pastes in ready-to-use form, which have already made their appearance in the market are sure to be popular. The problem can be of retaining plasticity and preventing fermentation; preservatives, if any, have to be permitted under the regulations for use in preparations having water. Use of pesticides in raising and storing spices has to be discreet so that right amount of residue of the approved pesticides are there.

Virtues

Apart from the sensory and culinary properties of spices, those of preservatives (antimicrobial), antioxidants and a few medicinal properties glorify the spices. Apart from the antioxidant property *in vitro* discovered earlier, the current knowledge of antioxidant action *in vivo* is very significant; the organic principles of the spices in general, in addition to capsaicin already known as such, appear to be able to act as physiological antioxidants. The mental fillip and excitement and euphoria generated by some spices are well-known. Other hedonic and medicinal virtues of different spices require to be investigated.

Useful Plants—What can be Achieved?

Though prevalent for a long time and good record of chemical research notwithstanding, the trade in plants and herbs has not been performing as desired. Steeped in poverty, the country should make good use of its resources. With all varieties of soil, climate, altitude and sunlight, India can cultivate any new plant and grow those already growing including those growing wild. One should know why India cannot avail of the full advantage on plants for medicine, food, food spices and other requisites.

Demand

It is heartening that only of late departmetns of forestry, health and science and technology of some governments have planned together to be more effective in the field of medicinal plants. Needless to say they are to do what they have not been doing successfully so far. As of now the relevant things are opportune. Demand for plant products is very high indeed not only for medicine but also for agricultural, industrial and household products. These are found and expected, to be non-toxic or less toxic and biodegradable or non-polluting. Any human activity or requirement can fall back upon plant products as raw material or basic starting substance. While it may sound like a recall to back to nature, it appears to be the smartest and wisest thing.

Plants provide us our foods and can possibly further do it for subsidiary foods and drinks. Some of the tribal foods not so far established as common food, can be investigated for general acceptance. Various drinks and decoctions, comparable to Ayurvedic tonics, can be made with herbs and medicinal plants; such drinks like RTS beverages are known to be popular in advanced countries specially with athletes and health conscious youth. Indonesia just of late began such use of herbal drinks. After all tea had a similar genesis and ginseng had a vitality craze. Some foods and drinks can thus be made, if necessary, with Ayurvedic ideas and formulations and some companies could make these popular here and abroad. It is a pity we could not take advantage of the Ayurvedic knowledge but promoted ginseng.

Production

Agriculture, as being practised now, should be modernised and should include more and more produce with introduction of newer plants and varieties.

Methods of biotechnology are known and are being applied variously. Such methods may be applied with medicinal plants and herbs too: e.g. manipulation of plant tissues in culture systems to regenerate plant through disorganised cell masses (callus) or to induce particular biochemical pathways to be operative. This can lead to plant improvement (separate from classical plant breeding), cloning of elite plants by micropropagation and creation of secondary metabolite profiles *viá* somaclonal variation by cells or organs *in vitro*. Micropropagation, regeneration, protoplast fusion and gene transfer (indirectly through *Agrobacterium* plasmid vector or directly DNA transfer by microinjection into cells or protoplasts, chemical or electrical permeabilisation and firing microprojectiles coated in DNA) and hairy root culture are a few methods of interest. Food and drug additives like colours, flavours and sweeteners and spices can be had from such cell, tissue and hairy root cultures.

Prospecting for drugs has been as old as chemistry itself. Laboratory work being more likeable, combinatorial chemistry came to be practised but soon pushed back in favour of plants. But not with whole plants but with isolated cells. The latter consume less cost (than whole plants) and are more manouverable. Thousands of specimens like microbial cultures can be maintained in cold and manipulated in various desired ways and screened (Anonymous, 1998). Plants are much more versatile in adaptation under stress, since they cannot run away, through synthesis of novel chemicals which can be utilised as pharma-,nutra-and phyto-ceuticals. Plant polyamines play vital roles in this regard (Chattopadhyay & Ghosh, 1998).

IPR

A central organisation can keep all data relating to such plants as relevant to India. For example, India's production antiquity, if available, discovery and knowledge, history of uses, known properties, technology of processing, products, etc. are to be known. Wherever possible patents should be taken as early as possible and countering done against biopiracy and infringement when necessary in India's IPR.

A good example which requires to be emulated had been that one NGO of Kerala drew from the knowledge of a tribe and then divided the accrued profit to the community.

We have not enacted on biodiversity and protection of plant though India ratified the convention on biodiversity four years ago. Attempts to plunder India's bioheritage and pirate the traditional knowledge of farmers and tribals will continue if we do not act.

India's cultural systems, emanating from the Vedas, encompass nearly 2000 plant species. There are nearly 8000 species mostly in the forest areas described in addition to Ayurveda by other traditional systems like Unani, Siddha and Tibetan (*Amchi*). More than 95% of medicinal plant species used by Indian industries is collected from the wild. In one estimate, the industries use more than 400 medicinal plants of which a mere 20 are commercially cultivated.

It is known that more than 40% of industrial population use traditional medicines to supplement modern ones. Even in 1987 *Rauwolfia serpentina* plant from India had a demand for 5000 MT as against a supply of only one fifty. Since then the demand for all medicinal plants and herbs rose several-fold with no idea of how much is being collected in the wild.

While activity in the wild plants is evident, the forest and welfare of the tribals come to the fore. They are even exploited on their expert knowledge.

Appropriate steps both nationally and internationally are overdue.

Technology

Value-added products rather than raw plants or their parts should be promoted for export, e.g. preserved, stabilised and dehydrated materials, different kinds of extracts, concentrates and isolates, oleoresins, food additives, subsidiary foods, tonics and health drinks. Research should be continued for new plant sources, new active chemicals, new technology and new products.

References

Anonymous, 1998. The Economist Newspaper, London, (Cited by Statesman 7June, 1998.)

Chattopadhyay M. K. and Ghosh, Bharati, 1998. Molecular Analysis of Polyamine Biosynthesis in Higher Plants. *Current Sc.* **74:** 517.

Recent Trends in Spices and Medicinal Plants Research pp. 78-82, 2000
Ed. Amit Krishna De
Associated Publishing Co., New Delhi

2.2

COFFEE

T. M. Das*

Abstract

Coffee is popular not only in India but all through the world. Some areas of cultivation, processing, chemistry & medicinal properties are discussed.

Introduction

Coffee plant belongs to the genus *Coffea* under the family Rubiaceae. The genus comprises 50-60 species of shrubs or small trees indigenous to tropical Asia and Africa. Among them 4 or 5 species are important as sources of commercial coffee. *C. arabica* is by far the most widely cultivated and is the source of Arabian coffee. It is a native of Ethiopia. *C. liberica* is the source of Liberian Coffee—a native of West Coast of Africa. *C. robusta* is the source of Congo coffee—a native of Congo. *C. liberica* and *C. robusta* is cultivated to a limited extent owing to its ability to thrive under varied climatic and soil situations and greater resistant to pest and diseases. There is another variety known as highland coffee of Sierra Leone obtained from *C. stenophylla,* but this species is not commercially important.

The commercial coffees of the world are grown mainly in tropical and sub-tropical countries, the chief areas lying between 25ºN and 25ºS. A little more than two-thirds of world's supply comes from Brazil and the rest from Ethiopia (Abyssinia), Arabia, Colombia, Costa Rica, Guatemala, Haiti, Hawaii, India, Jamaica, Java, Kenya, Mexico, Nicaragua, Philippine islands, Puerto Rico, Salvador, Srilanka, Sumatra and Venezuela.

In India, coffee is grown largely in south India on the foot hills of Western Ghats in Karnataka and sub-regions of Eastern Ghats in Tamil Nadu and Kerala. It is also produced in Andhra Pradesh. Other Asian countries famous for commercial coffee production are Srilanka, Arabia, Sumatra, Java and Philippine islands.

Coffee is grown mainly for seeds. The seeds are roasted and grounded into powder, which when soaked in boiling water yields a fragrant and stimulating beverage. The roasted coffee seeds contain 0.75-1.5% caffeine—the stimulating agent and a volatile oil—caffeol which is responsible for the flavour and aroma.

*Department of Agriculture, Calcutta University, 35 , B. C. Road, Calcutta–700 019.

The names given to the plant, its fruits, its seeds and the beverage prepared from these, are mostly derived from either of two words;—*'kahwah'* an Arabic term that originally denoted 'wine' and *'bun'* the Ethiopian name for the coffee plant of its beans. From these we have *cahwa, kawa, chaube, kapi, cave, kava, cafe, cafeier* and coffee, also *boun, bun, ban, ben, banu* and *buncha.* The earliest Arabic writers, however, used the Ethiopian name by itself or in combination, they called it *'buncho'* or *'Rhases bunco'.* Arabic writers described the coffee as a medicinal plant and attributed its origin to Ethiopia. They also knew that the fermented product of the plant was used as wine. But the technique of roasting of coffee bean and the preparation of a stimulating beverage out of it was unknown to them.

The real coffee plant would appear to have been satisfactorily established by botanists as indigenous to certain hilly regions of Ethiopia, of the Sudan, of Guinea and of Mozambique. The coffee is not mentioned in the *Koran,* nor of course is there any allusion to it in the Hebrew Scriptures. The people of Mecca, Medina and Bagdad apparently did not know of coffee till well into the 14th century of Christian era. Ibn Baithar, born at Malaga and who travelled during the 13th century in North Africa and Syria, makes no mention of coffee. The art of roasting the beans and preparing from a decoction was apparently a more recent discovery and one which may have been made in Paris. Prior to that, the *kahwah* that first attracted attention, was a preparation from the succulent rind or pulp of the coffee-cherry. This contains a fair amount of sugar, is often pleasantly enough flavoured and if a decoction made from it were allowed to stand for some short time it would for certain become alcoholic and might even be distilled into spirit.

Verzascha (Krautter-Buch., 1678, 179) gave a picture of roaster and described the preparation of a beverage from coffee seeds as well as from the husk. Sandys (Travels, 1610, 51) alludes to 'Coffee-Houses' of Constantinople. Herbert (Travels, 1677, 113, 311) speaks of 'Coffe' or 'Coho' as a Persian beverage prepared from the flower of the Bunny or Choava-berry and sold in shops. Coffee was known for centuries as a drug and many of the Arab writers speak of having been acquainted with its properties through the Persians. During the 15th Century, Abu Abdallah Muhammad Dhabbani Ibn of Aden had occasion to visit Persia and lastly Ethiopia. On that occasion he found the faithful partaking of coffee. Returning of Aden he introduced drinking of coffee among his followers. From Aden, however, the knowledge seems to have spread to Mecca, Medina and Cairo.

Difference of opinion exists regarding the first European who saw and described both the plant and the beverage. Clusius (*Arom. Hist.* 1574, 214-5) received a few coffee seeds from Dr. Alphonse Pancius of Ferrera during the summer of 1573 which he figured and described and mentioned its native name as *buna* or *elkave (al kave)* and that in Alexandria a drink was made from these seeds. Rauwolf visited Aleppo in November 1573 and saw the coffee plant, as also the beverage. He published his account in 1583 (*Beschreit. der Raiss* 1583, 103). Thus Clusius, not Rauwolf, as is commonly affirmed, should be viewed as the first botanist who examined and described the coffee berries. In 1615 coffee-drinking was carried to Venice. In 1644 Peter della Valle took it to Marseilles. John Houghton (*Phil. Trans.* London. 1809, iv., 420) says that Rastall, an English merchant, went to Leghorn, Italy in 1651 and there he found coffee-houses. In the year following Mr. Daniel Edwards, a merchant from Smyrna, Turkey, brought to England a Greek servant named Pasqua, who knew to make coffee. Shortly after, Pasque was able to set up a public coffee-house in Cornhill, England.

The history of the introduction of coffee into India is very obscure. Most writers agree that it was brought to Mysore some two centuries ago by a Muslim pilgrim named Babu Budan, who on his return from Mecca, brought seven coffee seeds with him. About the beginning of 19th century, there is however, no doubt that coffee had found its way to India. In 1823 a Charter was granted to Fort Gloster, near Calcutta to set up a coffee plantation. About the same time coffee was successfully grown in the Botanic Garden, Calcutta; but needless to mention that the industry of coffee planting nowhere found an abiding place on the plains of India but migrated to the hills of South India, in Mysore more specially, and thus into the very region where tradition affirms it had been introduced two centuries before. The first systematic coffee plantation was set up in India by Mr. Cannon near Chikmagulur in South India in 1830.

In Srilanka, it is believed coffee was introduced by the Arabs prior to the Portuguese invasion of that Island. It was commenced to be systematically cultivated by the Dutch from about 1690. In 1825 the first plantation was opened by an Englishman Sir Edwards Barnes. In 1877 it was estimated that the capital invested in coffee in Sri Lanka was close to £14,000,000.

Cultivation

Coffee grows best in a temperate climate in the tropical region. It requires a hot, humid climate with a fair interval of dry weather for ripening of the fruits. Coffee can be grown at all altitudes upto 6,000 ft. While *C. robusta* and *C. liberica* thrive well in regions of low altitudes, *C. arabica* gives the best results at high elevations from 2,500 to 5,000 ft. It needs at least 30-50 inches of rainfall, but can flourish in areas of high rainfall of 75-130 inches, provided the rain is well distributed throughout the year. Prolonged drought is detrimental. A temperature ranging from 12.5°–32°C is the most suitable for coffee cultivation. Coffee plants thrive best in a rich, well drained, friable, loamy soil containing abundant vegetable mould.

Coffee plant is a beautiful evergreen shrub or small tree attaining a height of 4.5 – 9.1 m. Leaves are opposite, stipulate, shining. Flowers are axillary in fascicles, white, fragrant, star-like, bisexual and epigynous. Fruits are small fleshy berries. Seeds are planoconvex, covered with a silver skin and enclosed in a dry husk-like parchment cover.

Coffee plants are grown from the seedlings which are raised from seeds in nursery beds. Healthy seeds are grown in January-March. Germination takes place within 6-8 weeks and they are transplanted and cultured in shaded, well manured nursery beds for about one year. In this stage, seedlings are regularly watered, weeded and sprayed with Borodeaux mixture. When seedlings become one year old they are transplanted in the pits of the permanent land during rainy season. The young plants are generally tied to thin stakes for support. Every year, manuring of the land with organic manure at the rate 5-10 tonnes per acre is essential. Besides, a fertilizer mixture containing 13.5-18.00 kg N, 18.0-27.0 kg $P_2 O_5$, 36.0 kg K_2O per acre are to be applied annually in two doses, before and after the monsoon respectively. Weeding and deep digging operations are done every year. Main stem of the plants is tipped twice, once when they attain a height of 0.62 m and the other one when they are about 1.5 m. The berries (fruits) begin to appear on the nodes of one year old branches.These one year old fruit bearing branches are pruned after the harvest. Generally, pruning is followed in the form of a single stem umbrella—in this method bushes tend to develop a dense upper canopy of crop bearing branches. Coffee plants begin to bear fruits to yield coffee in the third or fourth year and continues for about 40 years. The average yield of *C. arabica* is about 246.5-291.8 kg of clean coffee per acre while *C. robusta* yields nearly double the amount of *arabica*.

Processing

There are two methods of the processing of coffee for the market, e.g., the dry method and wet method. 'Cherry' or 'Native' coffee is obtained by the dry method while 'Plantation' or 'Parchment' coffee results from the wet method. In the dry method, fruits are spread on the floor exposing to the sun and dried thoroughly and uniformly. Next, the dried skin and pulp are cleaned off by peeling machine and the parchment is removed either by pounding in a morter or by any other mechanical means. The seeds or 'coffee beans, are then sorted by sieving into different grades, of which large round beans from single seeded fruits forming the top grade and packed in bags for marketing. In the wet method, the fruits are processed through a pulping machine, as a result the skin and part of the pulp are removed,—they are then placed in vats for fermenting. In this process, rest of the pulp ferments and washed off in water and finally seeds are dried in the sun or by artificial heat. After drying, the brittle parchment and the skin are peeled off by machines. The beans obtained by the above two methods are roasted. During roasting aroma, flavour and colour develop. Roasted beans are grounded to powder from which the drinking coffee is prepared. The broken, deformed and dark coloured beans constitute the lowest grade known as 'triage'.

Chemical Composition and Medicinal Property

Analysis of a large number of samples of coffee gave the following ranges of values: moisture 6-10%, nitrogen 1.6-2.3%, crude protein 10-14%, fat 10-14%, and ash 3.0-4.8% (Coffee Quality Investigation, Department of Agriculture, Mysore, 1947).

The average chemical composition of *C. arabica* beans is as follows: Raw leaves—moisture 8.98%, protein 9.87%, caffeine 1.08%, fat 12.60%, sugar 9.55%, dextrins 0.87%, caffetannic acid 8.46% and ash 3.74%. Roasted coffee—moisture 0.63%, protein 11.23%, caffeine 0.82%, fat 13.59%, sugar 0.43%, dextrins 1.24%, caffetannic acid 4.74% and ash 4.56%.

The caffeine content of the bean varies in different species. *C. arbica* contains 1.0-1.2%, *C. liberica*—1.4-1.6%, *C. robusta* 1.5-2.5% and *C. mauritania* 0.07%.

Seeds of the same species from different countries show varying amount of caffeine. The range of variation in the caffeine content of *C. arabica* beans is as follows: South India—1.0-1.2%, *Mocha* (Arabia) 0.6-0.9%, Brazil 0.5-0.9%, and Colombia—0.8-3.0%.

Raw and roasted coffee contain caffetannic acid which is known to be a mixture of chlorogenic acid and coffalic acid.

Sugars occur in raw beans mostly as glycosides. Sucrose, dextrin, cellulose, pentosans and hemicellulose are present.

Coffee bean oil and wax—Raw coffee bean contains about 10% oil and wax extractable with petroleum ether.

Recent Trends in Spices and Medicinal Plants Research pp. 83-85, 2000
Ed. Amit Krishna De
Associated Publishing Co., New Delhi

2.3

VANILLA—AN ORCHID SPICE FOR FLAVOUR

S. N. Potty* and V. Krishnakumar*

Abstract

India is known as the land of spices and has a long and glorious association with global spice trade as a major supplier right from ancient times. Indian spice industry was dominated by black pepper, popularly known as 'King of Spices' and small cardamom, the 'Queen of Spices'. Vanilla is the latest entry into the world of Indian spices and is the second costliest spice after saffron. Vanilla bean, the fruit of a climbing orchid, is the source of natural vanillin known for its pleasant flavour. It was Gobley who extracted vanillin from the beans which is now used in the preparation of ice creams, chocolates, bakery products, puddings, perfumes and liquors. Synthetic substitutes of vanillin are cheaper but they possess harsh and incomplete flavour while the taste of natural vanillin is smooth and highly acceptable to our tongue.

Vanilla, a native of Atlantic coast from Brazil to Mexico, is now grown on a plantation scale mainly in Madagaskar, Indonesia as well as Comoro and Reunion islands. The total world production of vanilla beans is estimated to be around 2500 tonnes valued at about Rs. 300 crores. Though, it occupies hardly 0.75% of the world import of spices in volume, in terms of value, it comes to 6 to 7 percent. The leading consumers of this high value low volume spice are the United States of America, France and Germany. Other significant importing countries are Switzerland, Canada, Japan, Denmark, the United Kingdom and the Netherlands.

There are over 50 species of vanilla, but only three are important as source of vanilla flavour. They are *Vanilla planifolia* Andrews, *V. tahitensis* J. W. Moore and *V. pompona* Scheide and among them *V. planifolia* is the most important and widely cultivated species.

No authentic information is available regarding the history of vanilla introduction into India. It is believed to be introduced to India in 1835 for the 'Spice Garden' in Kurtallam owned by the East India Company. Some pioneering work was done on cultivation practices and processing of the produce in the research station at Ambalavayal in Wynad and Kallar Burliar Fruit Research Station in Nilgiris. Encouraged by the success in vanilla cultivation in these research stations, a few planters in Wynad and in and around Mettupalayam, Tirunelvelli and Gudallore took

*Indian Cardamon Research Institute (Spices Board) Myladumpara, Idukki –685 553.

took up cultivation of this crop. Almost all those plants were damaged due to neglect and lost subsequently. Infact some of these vines have served as source of planting materials for the recent development of this crop in India. It is needless to mention that there has been a pause in vanilla development, but in the recent years Spices Board has come in a large way to develop this crop in our country.

This crop is suited for the moist humid tropics and subtropics and can be grown from sea level to 1500 m above mean sea level. It flourishes well under filtered light and hence is suitable as an intercrop in coconut and arecanut gardens. Well drained soils rich in organic matter is required for this crop. The propagation is mainly through shoot cuttings. Being a climber, the plants require support and are coiled around the branches or on horizontal supports. Vanilla requires a lot of organic matter for its growth and is generally free from major pests and diseases. They come to flowering in the third year and full bearing in 5 years. Due to peculiar nature of flowers, they require artificial pollination for the commercial production of beans. A special process of curing involving four stages namely sudden killing, sweating, slow drying and conditioning is needed for the development of this crop in Kerala, Tamil Nadu, Karnataka and parts of West Bengal and North Eastern states.

Vanilla comes up well in warm moist conditions as seen in humid tropics and grows in sandy loam to laterite soils with a gentle slope to prevent any water logging. Propagated mainly through vegetative shoot cuttings (60 to 100 cm long), vanilla requires a support for climbing and flourishes well when around 50% shade is provided. This allows the crops to be interplanted with coconut, arecanut or with shade trees. Of late, tissue culture plants are also being used as planting materials.

The plant is to be allowed to grow vertically only upto a height of about 1.5 m on the standards, later on the vines are to be coiled around the branches or on horizontal supports made parallel to the ground by connecting two vertical supports. Being a surface feeding orchid, vanilla requires a lot of organic matter for its growth. The plant is generally free of any disease, except the incidence of shoot tip rot caused by a fungus, which can be easily managed by prophylactic spraying of 1% Bordeaux Mixture before the commencement of monsoon rains.

If properly managed, vanilla comes to flowering in the third year of planting and due to the peculiar nature of the flower, requires artificial hand pollination for the commercial production of beans. The production reaches full bearing in the fifth year and on an average the annual production will be about 150 kg/ha of cured beans. The economic life span is around 15 to 17 years.

Harvesting of beans at the correct maturity and processing them by the right method with care and attention are very important for the proper development of flavouring constituents in the beans. Processing involves four stages namely killing of beans, raising temperature, slow drying and conditioning and the entire process takes about 2 to 3 months. The price of cured vanilla beans is determined based on the vanillin content and other quality parameters and will be around Rs. 1000 per kg.

Vanilla has a good scope for development in India considering the domestic demand and the export potential. Our present production is estimated at less than five tonnes and comes mainly from some plantations in Ernakulam and Wynad districts in Kerala as well as South Kanara district in Karnataka. Preliminary survey in our country indicates that vanilla can be commercially cultivated in parts of Kerala, Tamil Nadu, Karnataka, Lakshadweep, Andaman & Nicobar islands and certain North Eastern states. The Spices Board has launched many research and development programmes for vanilla cultivation in India. The Indian Cardamom Research Institute is presently implementing two schemes for vanilla cultivation which are funded by the Ministry of Science and Technology, Govt. of India.The Institute is also conducting regular training programmes on various aspects of vanilla cultivation for the benefit of planters.

Reference

Anonymous, 1992. Report of the National Committee on Vanilla. P. 6. Spices Board, Kochi.

Madhusoodanan, K. J. and Krishnakumar, V., 1994. Vanilla— an orchid spice crop. National Seminar on 'A. Decade of Orchid Research and Development, India International Centre, New Delhi, Oct. 24-25, 1994.

Purseglove J. W., Brown, E. G., Green, C. L. and Robbins, S. R. J., 1988. Vanilla. *Spices* **2**. 644.

Sasikumar, B., Rems, J. and Ravindran P. N., 1993. Vanilla—The much fancied export crop. *Indian Perfumer.* **37(4)** : 330-333.

Shadakshari, Y. G., Uthaiah B. C. and Chandrasekharaiah, S. C., 1994. A new method of hand pollination in vanilla (*V.planifolia* Andrews). *J. Spices & Aromatic Crops* **3(2)**:167.

Recent Trends in Spices and Medicinal Plants Research pp. 86-91, 2000
Ed. Amit Krishna De
Associated Publishing Co., New Delhi

2.4

COLLECTION, CLASSIFICATION AND EVALUATION OF ECOTYPES OF *BYADGI* CULTIVAR OF CHILLI IN DHARWAD DISTRICT OF KARNATAKA

V. A. Giritammannavar*, S. J. Patil* and N. V. Hiremath*

Abstract

Dharwad district of Karnataka has the distinction of having the highest average in India with over 1.30 lakh ha and the extensive cultivation of *Byadgi* cultivar known for past 50-60 years. *Byadgi* cultivar is widely grown in north Karnataka which produces high quality bright red fruits suited for export market. Owing to high natural cross pollination and introgression with other varieties/ hybrids, the genetic purity and quality of *Byadgi* cultivar has been deteriorated. An attempt was made to collect, classify and evaluate the major ecotypes of *Byadgi* cultivar to identify superior progenies. A total of 815 individual plant progenies from chilli fields in 13 taluks were collected during *Kharif* 1992. All the collections were systematically classified into two broad groups viz., thin *Kaddi* (K) and thick *Dabbi* (D) and further subgrouped as K1, K2, K3 and DI, D2 and D3 based on fruit features like length, shape, thickness, wrinkles, color, pedicel attachment, calyx growth and fruit yield per plant. Only 500 progenies were retained after evaluation of these collections during summer 1992-93 and 150 superior progenies were selected after evaluation during *Kharif* 1993. After further evaluation during summer 1993-94, 90 elite genotypes consisting of 30 progenies of *Byadgi kaddi*, 30 progenies of *Byadgi dabbi* and 30 progenies of intermediate were evaluated in randomized block design with three replications during *Kharif* 1994 at MRS, Dharwad and ARS, Hanamanamatti. Observations on 20 characters including fruit yield and capsaicin were subjected for genetic analysis. The number of fruits per plant, dry fruit weight, plant aspect, capsaicin and total yield/plant recorded high GCV, heritability and genetic advance, while days to 50% flowering and number of branches/plant registered low estimates of these genetic parameters in both locations. Significant and positive associations were observed between dry fruit weight with fruit number, average fruit weight, fruit length, number of seeds/fruit, capsaicin and seed weight/fruit. The genotypic path analysis revealed high direct positive effects of number of branches/plant, dry fruit weight on total yield/plant while fruit length and number of seeds / fruit showed direct negative effect in both locations. The results emphasized the need for inculsion of number of fruits per plant, dry fruit weight and plant aspect (uniform compact plant type) to improve the yield of *Byadgi kaddi* and *dabbi* cultivars.

Introduction

Chilli, also known as hot pepper or red pepper is the second largest commodity after black pepper in the international spice trade. Chilli is an indispensable spice in

*Department of Genetics and Plant Breeding, University of Agricultural Sciences (UAS), Dharwad-580 005,

Indian cuisine or recipes, hence it is the most common ingredient in daily diet as vegetable, condiment, culinary supplement or food additive for taste, colour and flavour. The chilli extract is widely used in pharmaceutical and processed food and beverage industries.

India ranks first in the world in area under chilli (9.18.000 ha) but production (7.89,000 t) and productivity (877 kg/ha) are considerably low (Annon,1994). The productivity levels in four major chilli producing states viz., Maharashtra, Karnataka, Andhra Pradesh and Tamil Nadu are also low. The high natural cross pollination from 40 to 50% puts the varietal identity at stake, even in the long known and popular cultivars like *Byadgi* in North Karnataka thus decreasing the yield potentials. The ever increasing population and enhanced demand for export have been exerting extra-pressure for high production necessitating concerted and systematic studies in the deteriorating material. Therefore, the present study was undertaken to investigate the variability in elite selections from *Byadgi* cultivar and to investigate the interrelationships among the 10 economic characters along with their relative contribution to yield, which are of immense use to effect the improvement in selection programmes.

Material and Methods

A total of 815 individual plant progenies from chilli fields in 13 taluks were collected during *Kharif* 1992. All the collections were systematically classified into 2 broad groups viz., thin *Kaddi* (K) and thick *Dabbi* (D) and further subgrouped as K1, K2, K3 and D1, D2, D3 based on fruit features like length, shape, thickness, wrinkles, color, pedicel attachment, calyx growth and fruit yield per plant. Only 500 progenies were retained after evaluation of these collections during summer 1992-93 and 150 superior progenies were selected after evaluation during *Kharif* 1993. After further evaluation during summer 1993-94, 90 elite genotypes consisting of 30 progenies of *Byadgi Kaddi*, 30 progenies of *Byadgi Dabbi* and 30 progenies of intermediate were evaluated in randomized block design with 3 replications during *Kharif* 1994 at MRS, Dharwad and ARS, Hanamanamatti.

Ten random plants in each replication were selected, tagged and used for recording the observations on ten characters (as listed in Table1) including total yield per plant. Capsaicin was estimated as per the procedure described by Sadashivam and Manickam (1992). The coefficient of variation, heritability and genetic advance were calculated following standard formulae of Burton (1952). Weber and Moorthy (1952) and Johnsen *et al.* (1995), respectively. Statistical analysis were performed by using the variances and covariances to estimate the phenotypic and genotypic correlations. The genotypic correlations were further partitioned into direct and indirect effects on total yield per plant by path analysis following Dewey and Lu (1959).

Results and Discussion

The data on genetic parameters of 90 elite genotypes for ten traits at two locations(Dharwad and Hanamanamatti) are presented in Table 1. The results indicated considerable range of variability for all the traits except days to 50% flowering.

However, most of the characters registered different coefficient of variation (GCV) values from Dharwad location to Hanamanamatti, indicating the variable effect of environments on the same genotype(s). The GCV for plant aspect, capsaicin, total yield and number of fruits were high, whereas low GCV values were recorded for days to 50% flowering and number of branches, in both the locations. Similar results were obtained by Arya and Saini (1977) but Gupta and Yadav (1984) reported high GCV for a number of branches/plant.

TABLE 1: GENETIC PARAMETERS OF 90 GENOTYPES OF *BYADGI* CHILLI FOR TEN TRAITS AT DHARWAD (D) AND HANAMANAMATTI (H) LOCATIONS.

Character (code)	*Location*	*Mean*	*Range*	*PCV (%)*	*GCV*	*h^2 (%)*	*Genetic Advance (GA)*	*GA an % of Mean*
1. Days to 50%	D	34.98	31.0-39.0	5.01	1.27	6.40	0.23	0.66
Flowering (FFL)	H	35.13	31.0-39.0	4.73	1.31	7.70	0.36	0.74
2. No. of branches	D	9.31	4.6-15.2	17.42	6.33	13.20	0.44	4.73
per plant (NBR)	H	9.95	6.4-19.8	17.43	5.88	11.40	0.41	4.12
3. No. of fruits	D	78.11	45.4-127.8	22.09	21.20	92.10	32.72	4.89
per plant (NFP)	H	79.32	41.2-114.4	19.11	17.07	79.70	24.90	31.39
4. Fruit Length	D	15.95	12.3-21.8	11.14	10.03	81.00	2.97	18.62
in cms (FLT)	H	14.86	8.4-18.3	10.97	7.00	40.70	1.37	9.22
5. Dry fruit wt.	D	1.58	0.99-2.26	15.78	11.14	49.00	0.26	16.46
in gm (DFW)	H	1.48	0.71-2.24	21.97	15.12	47.40	0.30	29.73
6. No. of Seeds	D	127.59	50.3-90.4	19.84	10.67	28.90	15.08	11.82
per fruit (NSF)	H	129.39	58.0-74.8	18.27	11.97	42.90	20.89	16.14
7. Plant aspect	D	2.47	1.0-5.0	32.26	30.11	72.90	1.31	53.06
(PAP)*	H	2.27	1.0-5.0	40.51	33.64	69.00	1.31	57.61
8. Capsaicin in	D	3.11	1.66-5.69	29.10	28.36	95.00	1.77	56.91
mg/100g (CPN)	H	3.01	1.58-5.59	29.83	23.99	64.70	1.22	39.61
9. Wt. of seeds/	D	0.66	0.38-1.06	17.53	11.18	40.70	0.10	15.17
fruit in gms	H	0.60	0.27-1.05	24.32	16.99	48.90	0.15	25.07
10. Total yield/	D	89.89	40.5-134.7	23.65	20.53	75.40	30.81	36.72
plant in gms (TYP)	H	88.17	41.2-117.5	21.41	16.38	58.50	21.47	25.81

* Scoring was done from 1 as best (uniform and compact plant type) to 5 as least (ununiform and variable plant type).

The heritability estimates ranged from 6.40% (days to 50% flowering at Dharwad) to 95.0% (capsaicin at Dharwad). Moderate to high heritabilites were observed for all the characters except days to 50% flowering, number of branches/plant and number of seeds/fruit in both the locations. The heritability estimates for total yield/plant, capsaicin and fruit length were low at Hanamanamatti than Dharwad thereby indicating the influence of environment on character expression. These results corroborate the findings of Gupta and Yadav (1984) and Bhagyalaxmi *et al.* (1980).

The estimates of heritablility alone do not give an indication about the amount of progress expected but genetic advance as percent of mean (GA%) acts as a measure of improvement by selection. The plant aspect and number of fruits/plant exhibited high GA% while days to 50% flowering and number of branches/plant revealed poor genetic advance upon selection. Arya and Saini (1977) and Bhagyalaxmi *et al.* (1990) had observed high GA% for total yield/plant but present study recorded moderate GA% indicating variable results, when different material is used for the study.

The Phenotypic Coefficients of Correlation (PCC) and Genotypic Coefficients of Correlation (GCC) for the 10 traits analyzed at Dharwad and Hanamanamatti, are presented in Table 2. Generally, GCC were higher than the PCC for most traits indicating the high heritable nature of characters studied. The association trends were almost same although nature and magnitude of correlation varied for the same character combinations. The PCC and GCC of total yield/plant were significant and positive with number of fruits/plant were significant and positive with number of fruits/plant and dry fruit weight in both the locations. The dry fruit weight had positive and significant relation with fruit length, number of seeds/fruit, capsaicin, weight of seeds/fruit at both locations. These findings are in conformity with the results of Rao and Chonkar (1981) and Bhyagyalaxmi *et al.* (1980).

TABLE 2: PHENOTYPIC AND GENOTYPIC CORRELATIONS AMONG 90 GENOTYPES OF *BYADGI* CHILLI FOR TEN TRAITS AT HANAMANAMATTI (UPPER DIAGNOL) AND DHARWAD (LOWER DIAGNOL)

Character	*P/G*	*1.FFL*	*2.NBR*	*3. NFP*	*4.FLT*	*5.DFW*	*6.NSF*	*7.PAP*	*8.CPN*	*9.WSD*	*10.TYP*
1. FFL	P	–	0.002	0.013	0.030	0.035	0.023	0.030	0.116	–0.013	–0.048
	G	–	0.267**	0.077	–0.352**	0.317**	–0.312**	0.119	0.077	0.486**	0.395**
2. NBR	P	0.063	–	0.121*	0.075	0.060	0.076	0.163**	0.025	–0.042	–0.086
	G	0.469**	–	0.352**	0.203**	0.126*	0.282**	0.338**	0.281**	0.069	0.279**
3. NFP	P	–0.072	0.002	–	–0.086	–0.087	–0.054	0.752**	0.078	–0.079	0.603**
	G	0.284**	0.314**	–	0.278**	–0.173**	–0.002	0.932**	0.044	–0.208**	0.693**
4. FLT	P	0.020	0.216**	0.253**	–	0.514**	0.961**	–0.064	0.293**	–0.198**	0.270**
	G	–0.058	0.601**	0.294**	–	0.843**	0.984**	0.211**	0.823**	0.128**	0.082
5.DFW	P	–0.018	–0.137*	–0.177**	0.531**	–	0.600**	0.046	0.222**	0.731**	0.234**
	G	0.164**	0.771**	–0.302**	0.418**	–	0.884**	0.150*	0.843**	0.412**	0.699**
6. NSF	P	0.002	–0.009	–0.125*	0.457**	0.455**	–	0.090	–0.346**	0.198**	0.344**
	G	–0.261**	0.194**	–0.273**	0.208**	–0.920**	–	0.235**	–0.106	–0..280**	0.144*
7. PAP	P	0.068	0.059	0.809**	–0.256**	0.145*	0.107	–	0.198**	0.147*	–0.113
	G	0.181**	0.144**	0.533**	0.320**	0.336**	0.301**	–	0.264**	0.294**	0.214**
8. CPN	P	0.067	0.105	0.094	0.624**	–0.364**	–0.316**	0.174**	–	–0.107	–0.46
	G	0.204**	0.245**	0.103	0.691**	0.451**	0.617**	0.200**	–	–0.367**	0.116
9. WSD	P	–0.049	–0.067	–0.143*	–0.475**	0.841**	0.532**	0.117	–0.351**	–	0.079
	G	0.546**	0.370**	–0.239**	0.292**	0.937**	0.923**	0.298**	–0.594**	–	0.347**
10. TYP	P	–0.042	–0.024	0.869**	0.025	0.219**	0.072	–0.767**	–0.036	0.193**	–
	G	0.212**	0.231**	0.949**	0.039	0.440**	0.048	0.399**	0.054	0.202**	–

* and **—Significant at 5% and 1% levels of probability, respectively.

P and G—indicate correlation coefficients at phenotypic and genotypic levels, respectively.

The capsaicin had negative and significant relation with weight of seeds/fruit while it exhibited positive association with fruit length and plant aspect in two locations studied. Similar results were obtained by Bhagyalaxmi *et al.* (1990) while Gupta and Yadav (1984) reported negative association of capsaicin with fruit length. The correlation coefficients had significant differences for days to 50% flowering, number of branches/plant, fruit length between two locations, and between phenotypic and genotypic level which may arise due to significant genotype × environment (G × E) interactions. The simultaneous improvement of two or more attributes is possible when positive correlation exists or proper balance between negatively associated traits is considered. In addition, the G × E interactions have to be kept at minimum by selecting genotypics with favourable traits after intensive evaluation over more locations. (Gupta and Yadav, 1984).

The path analysis for total yield per plant (TYP) was performed with nine independent traits at two locations and results have been compiled in Table 3. The individual character contribution to TYP has similar trend in both the locations, except few cases. The number of fruits/plant and dry fruit weight exerted maximum direct positive effects while fruit length and number of seeds/fruit had direct negative effects on TYP at both the locations. The magnitude of positive direct contribution to TYP through number of branches/plant and plant aspect was higher at Hanamanamatti than Dharwad. While negative direct effect on TYP was recorded through days to 50% flowering and capsaicin at Hanamanamatti.

TABLE 3: GENOTYPIC PATH OF NINE INDEPENDENT CHARACTERS ON TOTAL YIELD/PLANT AT DHARWAD (D) HANAMANAMATTI (H)

Character	*Location*	*1.FFL*	*2.NBR*	*3.NFP*	*4.FLT*	*5.DFW*	*6.NSF*	*7.PAP*	*8.CPN*	*9.WSD*	*Genotypic Correlation with TYP*
1. FFL	D	0.113	0.016	–0.134	–0.026	–0.150	0.187	0.019	0.018	0.117	0.212**
	H	–0.215	0.150	0.060	0.089	0.220	0.630	–0.521	0.619	0.513	0.395**
2. NBR	D	–0.053	0.134	–0.005	–0.265	0.239	0.542	0.015	0.022	0.080	0.231**
	H	–0.598	0.363	0.275	0.024	–0.201	0.595	0.741	0.438	–0.808	0.279**
3. NFP	D	0.032	0.070	1.103	–0.130	–0.275	0.196	–0.108	0.010	0.051	0.949**
	H	0.215	–0.198	0.882	0.018	–0.301	0.193	0.141	0.043	–0.199	0.693**
4. FLT	D	0.007	–0.020	0.345	–0.441	–0.741	0.652	–0.034	0.620	0.169	0.039**
	H	0.052	0.039	0.287	–0.429	–0.726	0.720	–0.097	0.106	0.136	0.082
5. DFW	D	0.079	–0.023	–0. 201	0.309	0. 810	–0.523	–0.042	–0. 048	–0.152	0.440**
	H	0.185	–0.089	–0.092	–0.185	0.745	–0.483	0.096	–0. 160	0.581	0.699**
6. NSF	D	0.029	–0.003	–0.320	0.400	0.928	–0.719	–0.034	–0.055	–0.220	0.048
	H	0.115	0.078	–0.234	0.356	0.879	–0.616	0.119	–0.152	–0.245	0.144**
7. PAP	D	–0.020	0.005	0.279	0.141	0.306	–0.246	0.107	–0.078	–0.094	0.399**
	H	–0.090	0.093	0.352	0.122	0.295	–0.374	0.218	–0.155	0.244	0.214**
8. CPN	D	–0.028	0.018	0.212	–0.265	–0.432	0.443	–0.021	0.089	0.128	0.054
	H	–0.065	0.157	0.132	0.198	0.451	0.404	–0.016	–0.104	0.256	0.116
9. WSD	D	0.061	–0.025	–0.280	0.348	0.852	–0.535	0.032	–0.053	–0.197	0.202**
	H	0.381	–0.077	–0. 216	0.255	0.794	–0.546	0.113	–0.127	–0.231	0.347**

*Residual (Dharwad) = 0.0617 Residual (Hanamanamatti) = 0.0492

**The diagonal values of the matrix are main effects (underlined values).

The indirect positive effects of number of branches/plant and fruit length through number of seeds/fruit; and number of seeds/fruit and weight of seeds/fruit through dry fruit weight were significantly high, revealing their importance in selection programmes. The fruit characters like length, number of seeds and wt. of seeds influenced the yield but through indirect effects of other characters thereby resulting in overall low or high GCC due to G $\times$ E interactions and camouflaging effect of such interactions on values of genotypes. These findings are in agreement with Rao and Chonkar (1981) and Depestre *et al.* (1988).

From the foregoing discussion, it may be concluded that number of fruits/plant, dry fruit weight and plant aspect should be given considerable emphasis in formulating selection criteria as they have high GCV, heritability, genetic advance, genotypic correlation and positive direct effect on total yield/plant in future chilli improvement programmes.

References

Anonymous, 1994. Performance of Agriculture in 1994. Centre for Monitoring Indian Economy, Bombay, India.

Arya, P. S. and Saini, S. S., 1977. Variability studies in pepper (*Capsicum* spp.) varieties. *Indian J. Hort.* **34** : 415-420.

Bhagyalaxmi, P. V., Ravishankar, C., Subramanya, B. and Babu, V. G., 1990. Study on heritability, genetic advance and character association in chilli. *South Indian Hort.* **38** : 15-17.

Burton, G. W., 1952. Quantitative inheritance in grasses. *Proc 6th Int. Grassland Congr.*, **1**: 277-283.

Depestre, T., Gomez, O. and Espinosa, J., 1988. Path coefficient analysis in sweet pepper, *Capsicum Newsletter* **7** : 37-38.

Dewey, D. R. and Lu, K. H., 1959. A correlation and path coefficient analysis of components of creasted wheat grass seed production. *Agron. J.*, **51**: 515-518.

Gupta, C. B. and Yadav, R. D. S., 1984. Genetic variability and path analyais in chilli. *Genetica Agraica*, **38:** 425- 432.

Johnsen, H. W., Robinson, H. E. and Comstock, R. S., 1955. Estimation of genetic and environmental variability in soyabean. *Agron*, J., **47:** 314-318.

Rao, P. V. and Chonkar, V. S., 1981. Correlation and path coefficient analysis in chilli. *Indian J. Agric. Sci.*, **52:** 857-860.

Sadashivam, S. and Manickam, A., 1992. Biochemical methods for Agricultural Sciences. Wiley Eastern Ltd., New Delhi, India.

Weber, C. R. and Moorthy, B. R., 1952. Heritable and nonheritable relationship and variability of oil content and agronomic traits in F2 generation of soyabean crosses. *Agron J.*, **44:** 202-209.

Recent Trends in Spices and Medicinal Plants Research pp. 92-95, 2000
Ed. Amit Krishna De
Associated Publishing Co., New Delhi

2.5

CERTAIN PRELIMINARY OBSERVATIONS ON THE MICROPROPAGATION OF THE ANTIDIABETIC HERB *GYMNEMA SYLVESTRE* R. BR.

T. Balakumar* and V. Selvakumar*

Abstract

A protocol for rapid *in vitro* propagation of the medicinal plant, *Gymnema sylvestre* R. Br. (Family Asclepiadaceae), widely prescribed as an antidiabetic herb, from single nodal bud explants is described. The explants collected from the actively growing climbers in our college herbal garden were surface decontaminated by adopting routine methods and were cultured on modified Murashige and Skoog (MS) medium containing 3% sucrose (w/v) and o.8% agar (w/v) and supplemented with various concentrations of BA + NNA, GA_3 + NAA and ADS + NAA. In general, the combination of BA and NAA was found to be more rapid and effective in the induction of bud breaking within a culture period of seven days, compared to the other two hormonal combinations tested. The frequency of bud breaking and shoot growth was proportional to the concentration of BA supplied in the medium. The highest frequency of axillary bud growth (90%) was observed with a hormonal treatment of 22.2 μmol BA+ 4.32 μmol NAA and further increase in BA concentration proportionately decreased the frequency of axillary shoot multiplication. Further, callusing at the base of the explants was inversely proportional to the increase in the concentration of BA. Nevertheless, with BA concentrations higher than 8.88 μmol (wherein the least callusing was recorded at the base of the explants) there was absolutely no growth of callus. Though the combination of GA_3 + NAA was also found to induce axillary bud growth in the explants, the highest frequency observed was only 40%. With the standardised *in vitro* culture conditions, the growing shoots are subcultures at intervals of fortnight duration. Further work is in progress in our lab to harden the plants raised through micropropagation techniques and to transfer them to the open field conditions.

Introduction

Gymnema sylvestre R. Br., popularly known as the "Australian Cow plant" (Sirukurinjan in Tamil) is a woody climber that belongs to the family Asclepiadaceae. The leaf material has been reported to stimulate insulin secretion. There are some

*Unit of Stress Physiology and Plant Biochemistry, Department of Botany, The American College, Madurai-625 002.

earlier reports that it has no component having direct action on the blood sugar level (*Congr. Luso-espan, Farm*, 1952, **3**,197; *Chem. abst.* 1954, 11728c). Nevertheless, it's leaves are used as an antidiabetic drug as they contain some active principles controlling blood sugar level (*Curr. Med. Pract.*, 1959, **3(5)**, 227; *Biol, Abstr.* 1959, 14336). The leaves contain gymnemic acid, the factor responsible for the property of paralysing the taste of sweet and bitter substances temporarily, when they are chewed.

Harvest of these plants on a mass scale from natural habitats and poor seed viability, led to the depletion of natural resources of this medicinal herb. This plant was distributed wildly in the Azhagar Hills (Eastern Ghats) located at 20 km from Madurai city. Now this plant is fast dissappearing from these ranges at an alarming rate due to biotic and abiotic pressure. For the conservation of these valuable medicinal resources micorpropagation would be of special use. Many such species have been propagated commercially by axillary bud proliferation as reported by Mantell *et al.* (1985), Pierik (1987) and Chu (1992). In the present study, we have attempted to establish a rapid protocol for the micopropagation of *Gymnema* starting from nodal explants.

Materials and Methods

Plants of *G. sylvestre* were collected from the college botanical garden wherein a germplasm resource of medicinal plants is maintained. The nodal segments were separated from the plant, defoliated and cut into small pieces of about 1.0-1.5 cm length. The nodal segments were initially washed in running tap water for one hour and then in mild soap solution (Labolene, Qualigens, India) for five minutes, and then washed with distilled water (5-8 changes) followed by a 30 sec dip in 80% ethanol containing streptomycin (20 ppm). After washing out the alcohol coating, surface disinfestation was done with 0.1% mercuric chloride solution (w/v) for 4-7 min depending upon the tenderness of the material. The first five nodes from the apical region were discarded. After repeated washing with sterile water, the explants were excised by trimming the cut ends and explanted on culture medium.

The basal medium (MS) used was that of Murashige and Skoog (1962) supplemented with 3% sucrose (w/v), 0.8% agar-agar, vitamin supplements such as nicotinic acid (10 mg L^{-1}), pyridioxine HCl (10 mg L^{-1}), thiamine HCl (0.1 mg L^{-1}), besides inositol (100 mg L^{-1}) and glycine (20-40 mg L^{-1}). Iron was supplied in the form of Na_2EDTA and $FeSO_4$. Shoot induction and multiplication was examined using MS medium variously supplemented with BA, GA_3, ADA and NAA. The nodal explants were inoculated in media supplemented with different concentrations of cytokinins with fixed concentration of NAA. Twelve replicates were tested for each treatment and the experiments were repeated at least thrice. The BA concentrations employed were: 2.22, 3.55, 13.32, 22.2 and 35.52 μmol also GA_3 at 2.32, 4.69, 6.98, 13.95, 23.25, 37.30 and 46.5 μmol with a fixed NAA concentration of 5.4 μmol. The pH of the medium was adjusted to 5.8 before autoclaving at 105 kPa for 20 min. The cultures were maintained at 25 ± 1°C with 55-60% RH and a 16 h PPFD supplied by cool white fluorescent tubes (35-50 μE M^{-2} s^{1}) and subcultured at ten day intervals on same media composition. Frequency of bud breaking, days required to bud break and also the callusing responses were recorded.

TABLE 1: INFLUENCE OF VARIOUS GROWTH REGULATORS TESTED ON BUD BREAKING IN *Gymnema Sylvestre* R.BR.

Treatment	BA + GA_3 µmol	*% of bud break*	*Days to bud break*	*Remarks*
0	NR	NR	NR	NR
NAA (4.32 µmol)				
+	3.55	NR	NR	callusing
BA	4.44	NR	NR	callusing
	6.66	13	12	
	8.88	26	12	
	17.26	33	10	
	22.20	80	6	
	26.64	67	6	
	31.08	60	7	
	35.52	60	7	
GA_3	5.78	NR	NR	callusing
	11.56	29	12	callusing
	14.45	33	12	
	17.73	40	8	
	23.12	40	8	
	28.90	33	10	

NR—No response

Results and Discussion

The morphogenic response of explants cultured on different concentrations of BA and Gibberellic acid are listed in Table 1. The culture medium devoid of growth regulators failed to stimulate bud break in the explants. However, addition of a low level of NAA (4.32 μmol) with BA(3.55 and 4.44 μmol) to the basal medium showed the friable callus (data not shown). Similar observations have also been made in *M. pruriens* (Chattopadhyay and Datta, 1995). Explants of the basal medium supplemented with 4.32 μmol of NAA and various concentrations of BA or GA_3 exhibited bud break responses. Depending on the concentration of cytokinin tested, the proportion of explants showing bud break varied from 15-80% and the number of days required for same was six to twelve, the maximum (85%) response was obtained when basal medium was supplemented with 22.2 μmol BA and 4.32 μmol of NAA and further increase in BA concentration proportionately decreased the frequency of axillary shoot multiplication. Though the combination of GA_3+NAA was also found to induce bud break in the range of 29-40%, the time required for bud break varied from 8-12 days, whereas in the combination of BA+ NAA it was only 6-12 days. The basal medium supplemented with GA_3 at 17.73 and 23.12 μmol showed a maximum bud breaking responses of 40% only.

The results obtained clearly indicate that for *in vitro* shoot induction in *G. sylvestre* BA would be an ideal hormone supplement compared to GA_3 (Table 1). Similar results with BA on micropropagation of apple and jack fruit have been recorded by Lundergan and Janick (1980) and Rahuman and Blake (1988). One of the major problems was failure of bud break response due to leakage of phenolic compounds from the explants as it is common with *in vitro* studies using Asclepiadaceae members. This was overcome by the addition of 0.3% activated charcoal in the medium. With the standardized *in vitro* culture protocol, the growing shoots are subcultured at fortnight intervals. Experiments for initiation of rooting are underway. Further work is in progress in our lab to harden the plants raised through micropropagation techniques and to transfer them to the open field.

Acknowledgement

The financial assistance from the Tamil Nadu State Council for Science and Technology, Chennai 600 025, under grant No. RJ/BS 02013 is gratefully acknowledged.

References

Chattopadyay, S., Datta, S. K., Mahato, S. B., 1995. Rapid micropropagation for *Mucuna pruriens F. pruriens* L. *Plant Cell Reports* **15**: 271-273.

Chu IYE., 1992. Perspectives of micropropagation industry. *In:* Kurata K., Kozai T., (eds.). *Transplant Production Systems.* Kluwer Academic, Amsterdam, pp: 137- 150.

Lundergan, C. A. and Janick, J., 1980. Regulation of apple shoot proliferation and growth *in vitro. Hort.Res* **20**: 19-24.

Mantell, S. H., Matthewa, J. A., Mc Kee, R. A., 1985. *Principles of Plant Biotechnology,* Blackwell Scientific, Boston, pp: 130-157.

Murashige, T. and Skoog, F., 1962. A revised medium for rapid growth and bioassays with tobacco tissue cultures. *Physiol. Plant.* **15**: 473-497.

Pierik, R. L. M., 1987. *In vitro* Culture of Higher Plants. *Martinus Nijhoff;* Dordrecht, pp: 183- 230.

Rahaman, M. A. and Blake, J.,1988. Factors affecting *in vitro* proliferation and rooting of shoots of jack fruit (*Atrocarpus heterophyllum* Lam). *Plant Cell Tiss. Org. Cult.* **13**: 179-187.

Recent Trends in Spices and Medicinal Plants Research pp. 96-99, 2000
Ed. Amit Krishna De
Associated Publishing Co., New Delhi

2.6

IN VITRO INDUCTION OF AXILLARY BRANCHING IN THE MEDICINAL PLANT *Acalypha fruiticosa* Forsk.

V. Selvakumar* and T. Balakumar*

Abstract

Acalypha fruiticosa **Forsk. (Family: Euphorbiaceae) is a shrub whose leaves serve as a good source of medicine for many stomach and intestinal disorders. This plant is distributed wild in the Azhagar Hills (Eastern Ghats-10° 0' to 10°30' N & 75°55' to 78°20' E) located at 20 km from Madurai city. A comparison of floristic surveys conducted in 1980 and 1995 revealed an alarming situation that nearly 25 plant species in these hills disappearing from this area. We have made certain attempts to conserve these local medicinal plants by raising a germ plasm bank in our college botanical garden. We are also trying to propagate these plants by adopting both conventional and micropropagation methods. In this paper, we present a standardized protocol for the rapid induction of axillary branching in *Acalypha fruiticosa* single nodal segments. Micropropagation was achieved through forced axillary branching on modified Murashige and Skoog's (MS) medium supplemented with 37.2 μM kinetin + 5.4 μM NAA and 22.2 μM BA + 5.4 μM NAA, with 3% sucrose and 0.8% agar-agar. Although bud breaking was noticed in both the treatments tested on the fourth day of inoculation, the kinetin + NAA combination registered the highest frequency of axillary branching (75%) compared to the BA + NAA treatment. Our experiments are underway to induce rooting in these axillary branches also for making them suitable for field conditions.**

Introduction

Medicinal plants continue to provide healthy security to rural people in primary health care. According to the WHO, over 80% of people in developing countries depend on traditional medicine for their health needs. In India, modern health coverage of rural populations varies from 3 to 30%, which means that between 400 and 500 million people take recourse to some form of traditional medicine as the only feasible alternative. This is also borne out by the fact that there still exist over 10 lakh (one million) traditional village-based carriers of the herbal medicine traditions in the country. A comparison of floristic surveys conducted in 1980 and 1995 in the Azhagar Hills (Eastern Ghats) of Madurai revealed an alarming situation that the maximum threats to the existence of medicinal plants are man-made habitat loss,

*Unit of Stress Physiology and Plant Biochemistry, Department of Botany, The American College, Madurai 625 002.

fragmentation and degradation due to population and developmental pressures are the bane of biodiversity. What threatens medicinal plants the most is over exploitation from the wild for commerce (WHO, ICUN and WWF: GLAND, 1993). So we are trying to propagate those plants by adopting both conventional and micropropagation methods in our laboratory. In this paper, we present a standardized protocol for the rapid induction of axillary branching in *Acalypha fruticosa* using single nodal segments. *Acalypha fruticosa* Forsk, a member of Euphorbiaceae, is a shrub commonly found in South India.

The leaves are of medicinal importance and are prescribed for stomach disorder, dyspepsia, alliterative and attenuant. It is generally propagated by nodal segments, because seeds are not available at all seasons, besides its seeds are of small size and loses their viability on storage. Therefore, an attempt has been made to develop an economically cheaper micropropagation technique.

Materials and Methods

Plant material: Plants of *Acalypha fruticosa* with actively growing shoots collected from the Azhagar Hills provided the explant source. Healthy nodal explants (0.8—2.0 cm) with a single dormant axillary bud were used as explants which were washed thoroughly under running tap water for 1 to 2 hours, followed by immersion in a 5% labolene solution (Qualigens, India), in distilled water for five minutes and then under running tap water for 30 minutes followed by a fungicide treatment (Bavistin-Carbendazim 50% wp). Then the explants were surface sterilized with 0.01% (w/v) aquous mercuric chloride for five minutes and rinsed several times with sterile distilled water. Each nodal segments was cut to 1.5 cm long and placed *in vitro* on MS medium (Murashige and Skoog, 1962) supplemented with 0.2-1.0 mg BA or KN or ADS and GA_3 with NAA (0.2 mg L^{-1}) in various combinations. The pH of the medium was adjusted to 5.6 to 5.8 before gelling it with 0.8% (w/v) agar. After melting 0.8% agar (Qualigens), the media were dispensed into culture vessels (20 ml media per 150 x 25 mm tube). All the media were autoclaved for 20 min. at 121°C (108 kPa). Cultures were maintained at 25 ± 1°C with 16/8 hrs photoperiod provided by a set of white fluorescent tubes (33-50 $\mu Em^{-2}\ s^{-1}$) with 55-60%RH. Eight to ten replications were tested for each treatment and experiments were repeated at least thrice.

Results and Discussion

Explants were collected from a plant growing wild during October to February to determine the most appropriate time to initiate cultures (vegetative period). The serious problem of contamination experienced when the explants were derived from juvenile and mature explants, which was overcome by treatment with streptomycin in 80% ethanol. The morphogenic responses of nodal explants were observed on MS medium supplemented with 3% sucrose, 37.2 μmol KN + 5.4 μmol NAA and 22.2 μmol BA + 5.4 μmol NAA. Although bud breaking was noticed in both the treatments on the eighth day of inoculation, the combinations of KN with NAA registered the highest frequency of axillary branching (75%) compared to the BA with NAA treatment (50%). The combination of 22.2 μmol BA and 5.4 μmol NAA required ten days for bud breaking, whereas the combination of 37.2 μmol KN with 5.4 μmol of NAA required only eight days and also the frequency of bud breaking was higher in KN + NAA treatment compared with BA + NAA (Table 1.) Callusing at the cut end of the explants was observed in lower concentrations of the both BA and KN in combination with NAA. Initially the callus was yellowish green and later on turned dark brown in two weeks. Similar observations have been reported in *Pterocarpus santalinus* L. (Sita, 1992) and *Morus alba* (Bhojappa *et al.* 1994). From each developing bud, a single shoot elongated within eight to twelve days (data not shown). Pattnaik *et al.* (1996) have reported similar response in *Morus australis.*

TABLE 1: INFLUENCE OF VARIOUS GROWTH REGULATORS ON BUD BREAKING IN *Acalypha fruiticosa* Forsk.

Treatment	*BA/KN μmol*	*% of bud break*	*Days to bud break*	*Remarks*
0	NR	NR	NR	NR
NAA (5.4 μMOL)				
+	2.22	NR	NR	callusing
BA	3.55	NR	NR	callusing
	13.32	25	12	callusing
	22.32	50	10	
	35.52	33	12	
KN	2.32	NR	NR	callusing
	4.69	NR	NR	callusing
	6.98	16	12	
	13.95	42	10	
	23.25	58	8	
	37.20	75	8	
	46.50	33	10	

NR—No response

The nodal segments were more responsive than the apical shoot buds. This was the case with *Pistacia vera* also (Barghchi and Alderson, 1982). Such differential morphogenic responses could be due to differences in the physiological status of the buds on different regions of a stem (Vieitez *et al.* 1985). Experiments are underway to induce rooting in these axillary branches and for making them suitable for transfer to field conditions.

Acknowledgement

The financial assistance from the Tamil Nadu State Council for Science and Technology, Chennai 600 025, under grant No: RJ/BS 02013 is gratefully acknowledged.

References

Barghchi, M., Alderson, P. G., 1982. *Acta Hort.* **131**:1-6, 1982.

Bhojappa, G., Sha Vallikhan, P. S., Sudarsanam, G. and Rao, D. R., 1994. *J. Phytological Res.* **7**: 1-7.

—1983. Guidelines on the conservation of medicinal plants: WHO, IUCN & WWF: GLAND, Switzerland.

Murashige, T. and Skoog, F., 1962. A revised medium for rapid growth and bioassays with tobacco tissue cultures. *Physiol. Plant.* **15**: 47-497.

Pattnaik, S. K., Sahoo, Y. and Chand, P. K., 1996. Micropropagation of a fruit tree, *Morus australis* Poir. Syn. *M. acidosa* Griff. *Plant Cell Reports* **15**: 841-845.

Sita, G. L., 1992. *Curr Science* **62**: 532-535.

Vieitez, A. M., Garmen, S. M., Vieitez, E., 1985. *J. Hort Sci* **60**: 99-106.

Recent Trends in Spices and Medicinal Plants Research pp. 100-105, 2000
Ed. Amit Krishna De
Associated Publishing Co., New Delhi

2.7

IN VITRO PROPAGATION OF *Scoparia dulcis* LINN., A MEDICINAL HERB

G. Muthuraman*, P. Brinda* and T. S. Lokeswari**

ABSTRACT

Commonly known as sweet broom or goat weed, *Scoparia dulcis* is a perennial herb used for ailments such as fever, dysentry, hypertension and others. Tissue culture studies have shown that regenerating shoot cultures had chemical yields that was 60% of the *in vivo* tissues. Micropropagation in this plant has not however been reported. In the present report, shoots were produced from nodal explants in the presence of benzyladenine (BA) and thidiazuron (TDZ). With BA shoot clusters were observed while in thidiazuron at one tenth the concentration of BA, normal length of the shoots were produced. Shoots were rooted in basal medium and successfully acclimatized for field trials. The horticultural conditions for growing this crop was also achieved. Thus, *in vitro* multiplication in this herb is reported for the first time from an Indian population of the SDX chemotype.

Introduction

A member of Scrophulariaceae, *Scoparia dulcis* was introduced into India from America. This perennial herb, commonly called goat weed or sweet broom weed (*Santal; Jastimadhu: Chakrai vampoo*) became a weed spreading along rice fields in S. India, the tea plantations of Assam and Bengal (Haridas and Sharma 1973; Anonymous 1977; Barthakur *et al.*, 1989) and citrus orchards (Megh Singh, *et al.*, 1986). Three chemotypes of the plant have been identified based on the predominant type of diterpene produced in the plant.

Chemotype	*Predominant diterpenes*	*Distribution*
SA/SDB type	Scoparic acids A, B & C	Paraguyan
SDB	Scopadulcic acid B	Paraguyan/Formosan
SDX	Scopadulciol, scopadiol	Formosan

The folklore usage of this plant is reported from Taiwan, Paraguay, Brazil, Bangladesh and India. It is widely used in treatment of ailments such as fever,

* Captain Srinivasa Moorthy Drug Research Center, Anna Hospital Campus, Arumbakkam, Chennai 600106.
** SPIC Science Foundation, Centre for Biotechnology, 111 Mount Road, Guindy, Chennai 600 032.

emetic, cough diarrhoea, bronchitis, hypertension, toothache, diabetes, diuretic, stomach disorders, gravel, kidney complaints and wounds (Anonymous, 1988; Chow *et al.,* 1974, Penna, 1964; Kirtikar and Basu, 1984;). Several novel biologically active compounds in this herb was demonstrated (Table 1; Hayashi *et al.,* 1993, 1994, 1996; Rastogi and Mehrotra, 1993). While this herb is difficult to control in plantations using known herbicides (Megh Singh, *et al.,* 1986), poor germinability of the seeds in temperate climates has been reported (Jain and Singh, 1989).

TABLE 1: BIOLOGICAL ACTIVITY OF NATURAL COMPOUNDS FROM *Scoparia dulcis.*

Ailment	*Compound*	*Biological Evidence*
Anti diabetic	Amellin (Patent : 1948)	Relieves symptoms of glycosuria, hyperglycemia
Antiviral	Scopadulcic acid B (SDB)	Inhibition of Herpes Simplex Virus type I
Anti-ulcer agent	Scopadulciol (SDY)	Inhibits H^+/K^+ ATPase of gastric vesicles
Inhibits insects; antifungal	6-methoxy-2 benzoxazolinone	Inhibits European corn borer; aphids; *Helminthosporium; Fusarium* & *Gibberella* spp.
Hepatosis	Scoparic acid A (SA)	Inhibition of β-glucuronidase
Anti-tumour	Scopadulcic acid B (SDB)	Cytotoxic & inhibits Ehrlich ascites tumour

Hayashi *et al.*, (1993; 1994 & 1996) reported regeneration of the plants from calli. The chemical yields in regenerated shoots were found to be in the range of 2-3% of parental tissues. This could be enhanced to 60% of parental tissues in regenerated shoots from a transformed line (Yamazaki *et al.,* 1996). Micropropagation protocols were however not optimized for any of the chemotypes. Our data presents detailed *in vitro* responses in the herb prevalent in South India.

Materials and Methods

Plant Material

Plants and seeds from wildly growing populations of *Scoparia dulcis* were collected from Porur and Guindy, Chennai and grown in pots in the greenhouse. Seeds or explants were surface sterilized with 0.1% mercuric chloride for 5 or 10 min respectively before initiating on a basal medium.

Culture Media and Growth Conditions

All media contained Murashige and Skoog's basal (1962) composition amended with growth regulators for the different experiments. The cultures were incubated at 25 $\pm$ 2 °C with a photoperiod of six hrs light and intensity of 2000 lux.

Cultivation

The seedlings of seeds collected from the wild were used as propagules. The plot (size two cents) was ploughed well with farmyard manure and irrigated before commencing planting operations. Plants or seeds were planted at half foot distance with half foot ridge between rows.

Experiments

Initiation

Various explants of this herb were initiated on a basal medium. At the end of four weeks, the *in vitro* response was assessed in terms of %sterile cultures and shoot response.

Multiplication of Shoots

Multiple shoot induction from nodes and shoot tips was evaluated in response to two growth regulators benzyl adenine (BA, 1.0 mg/L) and thidiazuron (TDZ, 0.01 mg/L). Each bottle was inoculated with ten single nodes or shoot tip explants. Four weeks after incubation the number of multiple shoots per node and shoot length was noted.

Biomass Increase in Shoot Cultures

Shoots (1.5 g fr.wt) grown on MS basal medium were used as inoculum for this experiment. In this experiment growth of cultures was compared in agar and liquid media (MS basal) amended with growth regulators -BS (1 or 2 mg/L) or TDZ (0.01 or 0.5 mg/L). Cultures were grown in jam bottles and incubated static for agar and rotated at 150 rpm in a shaker for liquid cultures (60 ml.) Fresh weight determined at the end of four weeks was used to compute increase in biomass of this herb. Five replicates per medium were tested.

Acclimatization and Cultivation

Shoots obtained from *in vitro* cultures were rooted on basal medium containing 0.75% sucrose. The plants were acclimatized in pots containing vermiculite and sand. Pots were covered in plastic bags to maintain high humidity.

Results and Discussion

Initiation Response

The results presented in Table 2 showed that except for the internodes (40.5%) the percentage of sterile cultures established ranged from 76-88%. All the four different

TABLE 2: RESPONSE OF DIFFERENT EXPLANTS DURING INITIATION

Type of Explant	*Sterility %*	*Shooting Response*	*No. of Shoots/ Explant*	*Height (cm) Range*
Node	76.25 (4.2)	75 (4.08)	1–3	3.5–8
Internode	40.5 (4.7)	31.25 (3.4)	3–12	0.5–3.5
Leaf	88.75 (3.43)	23.75 (1.6)	3–20	1–5.5
Shoot Tip	87.5 (3.19)	87.5 (3.19)	1	8

explants produced shoots with different frequencies. Shoots were induced maximally in node and shoot tip explants (75 & 87.5%) in a basal salt medium. Multiple shoots were generally seen in node, internode and leaf explants. The shoot tips showed a mere rapid elongation from 1-8 cm in height.

Multiplication of Shoots

The results in Table 3 demonstrate that the number of shoots induced per node in the presence of thidiazuron was on an average 2.5 shoots/node (opposite). In the presence of BA, the shoots were numerous and short implying an organogenic response to this regulator.

TABLE 3: MULTIPLE SHOOT FORMATION FROM SINGLE NODES/SHOOT TIPS.

	TDZ			*BAP*		
Explant	*Multiple shoots (Avg.)*	*Multipli-cation Rate*	*Shoot Length (cm) Range*	*Multiple Shoots (Avg,)*	*Multipli-cation Rate*	*Length (cm)*
Node	26.2	2.58	2.7–5.1	numerous	high	0.5–1.1
Shoot Tip	27.4	2.7	1.1–6.4	numerous	high	0.4–1.8

Biomass Increase in Shoot Cultures

Biomass increase of the shoots was at least twice more than the basal medium when calli were grown with BA (1mg/L) or TDZ (0.5 mg/L) in agar cultures (Fig. 1). In a liquid culture system all growth regulators improved biomass by two times over the basal medium while TDZ at 0.01 mg/L showed only a marginal increment. This reflects the insensitivity of the tissues to such a low concentration of the added growth regulator.

Acclimatization and Cultivation

The *in vitro* shoots could be rooted in four weeks on a basal medium (100%). The hardenend plants were transferred to the field with no loss in survival. Some of the plants when incubated in test tubes produced flowers *in vitro* (in 35 days). These self fertilized seeds appeared normal and could be germinated (100%). Table 4 gives the cultivation parameters for this medicinal crop. At a seeding rate of 1kg the biomass yield for the crop is 15 times in 120 days of cultivation. Chemical yield (3.391g/kg d.wt.) was computed from an estimate of the diterpene, scopadulciol, *in vivo* tissues (Prof. Hayash, T.).

TABLE 4: CULTIVATION OF *SCOPARIA DULCIS*

Seedling Rate/cent	1 kg
Yield/cent	15 kg or 1500 kg/ac
Cultivation cost/cent	Rs 280
No. of days (atleast 3 crops possible per soil amendment)	120 d/crop
Chemical yield (scopadulciol)	3.391 g/kg d. wt

Thus induction of shoots with BA was prolific and this was similar to the earlier report (Hayashi *et al.*, 1994). Thidiazuron at 1/10th amounts of benzyladenine,

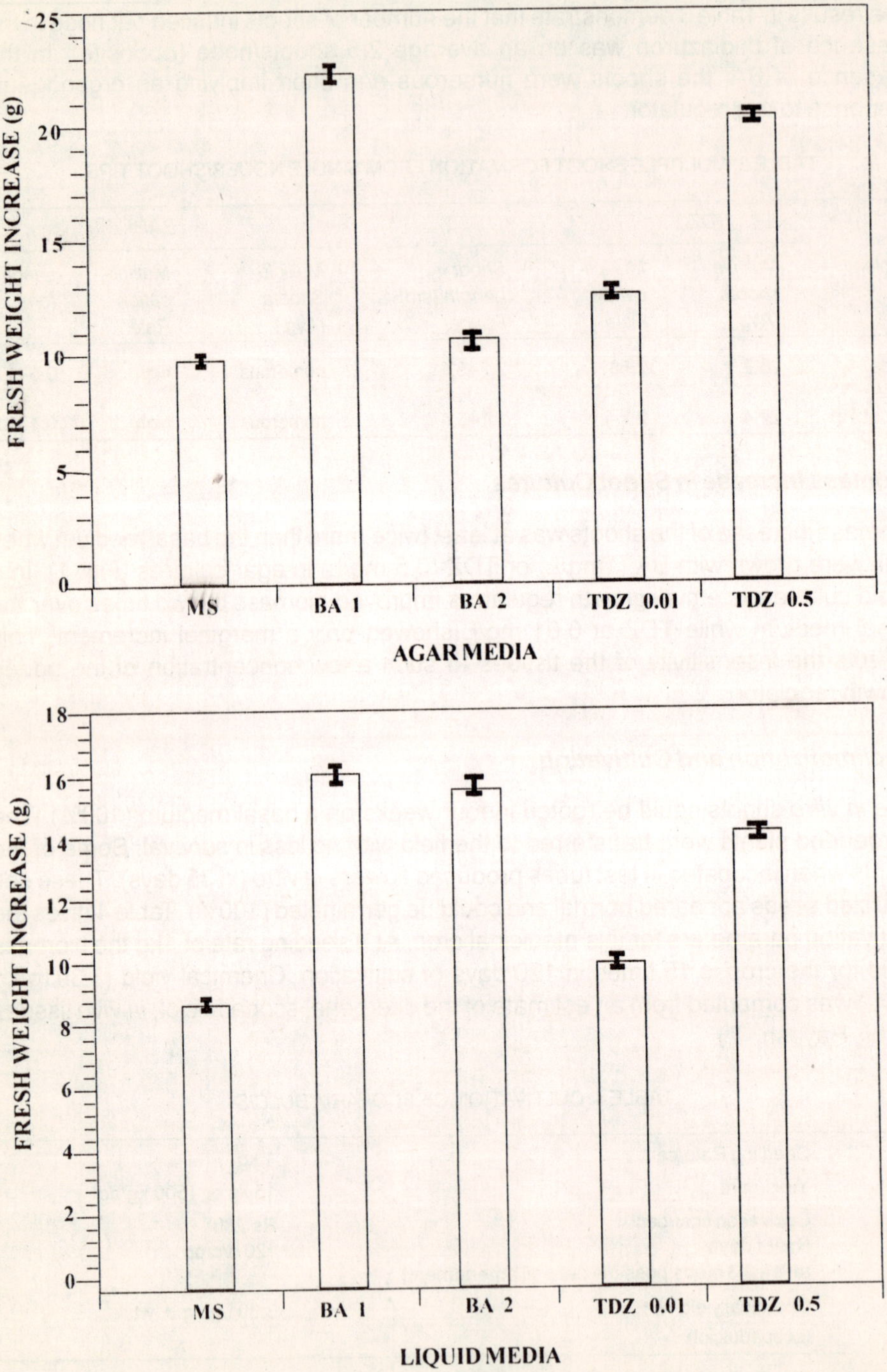

Fig. 1. Mean Fresh Weight Increase of Shoots. All Media are Based on MS Basal with BA or TDZ (mg/1).

induced normal sized shoots which is reported here for the first time. All shoots obtained from *in vitro* culture could be hardened and grown in pots. The chemical yields of *in vitro* grown shoots was about 5% of the parent tissues (data not shown) and agreed with reported values.

Acknowledgements

We wish to thank Ms Sujatha Jose, R. Gandhimathi, V. Sayee and A. Bhagyalakshmi for their team support and help. We specially acknowledge assistance from Ms P. K. Salama, who cultured the tissues for the work. GM is specially thankful to Dr George Thomas, Director, SPIC Science Foundation, for permission and facilities to carry out this work.

References

Anonymous, 1988. *Wealth of India*, Raw *Materials* Vol. **IX**, CSIR, New Delhi, pp. 260-261.

Anonymous,1977. Annual Scientific Report. Tocklai Experimental Station. pp. 19-20.

Barthakur, B.K., Dutta, P. and Begum, R.,1989. Vesicular Arbuscular Mycorrhiza (VAM) in some weed species of tea fields. *Tea and a bud,* **36**:8-9.

Chow, S. Y., Chen, S.M., Yang, C.M., Hsu, H., 1974. Pharmacological studies on Chinese herbs. 1. Hypotensive effects of 30 Chinese herbs. *J. Formos. Med. Assoc.* **73**: 729-739.

Haridas, P., Sharms, V. S., 1973. Some common weeds of South Indian Tea fields. 12. *Scoparia dulcis* and *Stachytarpheta urticaefolia* (salish.) *Sims. Planters-Chronicle,* **68**: 495-496.

Hayashi, T., Okamura, K., Kawasaki, M. and Morita. N., 1993. Production of diterpenoids by cultured cells from two chemotypes of *Scoparia dulcis. Phytochemistry,* **33**: 353-356.

Hayashi, T., Gotoh, K., Ohnishi, K., Okamura, K. and Asamizu, T., 1994. 6- Methoxy-2 benzoxazolinone in *Scoparia dulcis* and its production by culture tissues. *Phytochemistry,* **37**: 1611-1614.

Hayashi, T., Gotoh, K . and Kasahara, K., 1996. Production of scopadulciol by cultured tissues of *Scoparia dulcis. Phytochemistry,* **41**: 193-196.

Jain, R. and Singh, M., 1989. Factors affecting goatweed (*Scoparia dulcis*) seed germination. *Weed Science,* **37**: 766-770.

Kirtikar, K. R. and Basu, B. D., 1984. *Indian Medicinal Plants.* Bishen Singh Mahendra Pal Singh, Dehra Dun 1823p.

Murashige, T., Skoog, F., 1962. A revised medium for rapid growth and bioassays with tobacco tissue cultures. *Physiol. Plant.*, **15**: 473-479.

Megh Singh, D., Tucker, P. H., Lowery, C. A., 1986. An evaluation of herbicides for weed control in a young citrus grove. *Proc. Florida State Hort. Soc.,* **98**:14-16.

Penna, M., 1964. *Dicionarios brasilereros de plantas Medicinas*, Kosmos, Rio de Janeiro, Brazil, pp 351-352.

Rastogi, R. P. and Mehrotra, B. N., 1993. *Compendium of Indian Medicinal Plants.* Vol **1-4** CDRI, Lucknow and PID, New Delhi.

Yamazaki, M., Son, L., Hayashi, T., Morita, N., Asamura, T., Mourakoshi, I. and Saito, K.,1996. Transgenic fertile *Scoparia dulcis* L folk medicinal plant conferred with a herbicide resistant trait using an R1 binary vector. *Plant Cell Rep.* **15**: 317-321.

Recent Trends in Spices and Medicinal Plants Research pp. 106-121, 2000
Ed. Amit Krishna De
Associated Publishing Co., New Delhi

2.8

ROLE OF BIOTECHNOLOGY IN THE IMPROVEMENT OF BIOACTIVE COMPOUND WITH SPECIAL REFERENCE TO THE MEDICINAL PLANTS

Amita Pal* and Sangita Basu*

Abstract

Medicinal plants are the biological wealth of the nation. Plant biotechnology can play a key role in the conservation of endangered medicinal plants or to sustain the productivity of the biologically active compounds under *in vitro* condition. Appreciation of this technology stems from the premise that a number of factors (biological, chemical and physical) can be manipulated simultaneously *in vitro* to enhance the yield of bioactive compounds. *In vitro* culture technique often brings about 'metabolic surprises' and new variables with efficient drug potential can be recovered. The 'metabolic surprise' may be explained by the existence of controlling elements as envisaged by Barbara Mc Clintok. Activity of such controlling elements may be triggered in the cultural environment. Insertion of controlling elements may suppress the transcriptional activity at the site of insertion or may produce aberrant transcript(s), which may be turned out to be a desired compound. During the last decade of this century medicinal plants have received attention from molecular biologists and attempts are being made to transfer specific genes whose expression is predicted to alter secondary metabolism using recombinant DNA technology. Thus, manipulation of secondary product accumulation in transgenic plants has become feasible.

Introduction

Medicinal compounds are abundant in nature. They are present in plants microorganisms, marine-organisms, arthropods and other animals. These bioactive compounds are secondary metabolites, i.e. derivatives of primary metabolites. Apparently, they do not take part in any of the fundamental metabolic processes in plants and as such, are referred to as secondary metabolites. However, these are no less important than primary metabolites and play a vital role in medicine as analgesics (morphine), anti-tussives (codeine), antihypertensives (reserpine), cardiotonics (digoxin), anticancer agents (vinblastine, taxol), anti-malarial agent (artimisin, quinine), antiulcer compound (phenylpropanoids) and anti-HIV (castanospermine) etc. In this review article we have dealt only with bioactive compounds that are present in higher plant taxa. Generally, these bioactive compounds

* PMCG Section, Bose Institute, P 1/12 CIT Scheme VII M, Kankurgachi, Calcutta-700 054

are produced in very low quantity in plants, however, as the demands of bioactive compounds are very high, such chemicals are extremely high priced. A list of important medicinal plants and their active principles are depicted in Table 1.

TABLE 1: NAME OF THE SOURCE PLANT AND CHEMICAL SUBSTANCES ACTIVE AGAINST FEW MAJOR DISEASES OF PRESENT DECADE.

Plant	*Isolated Compound*	*Physiological Activity*
Artimissia annua	Artimisin (diterpene)	Anti-malarial
Azadirachta indica	Azadirachtin	Blood-purifier, Anti-leprosy
Berberis parvifolia	Berberine (glycosidase inhibitor)	Anti-neoplastic
Catharanthus roseus	Vincristrin Vinblastine	Antileukaemic
Coreopsis sp.	Phenylpropanoids	Anti-ulcer
Dioscorea deltoidea	Diosgenin	Steroid, contraceptives, Sex hormone precursor
Rauwolfia serpentina	Reserpine	Anti-hypertensive
Sapium sebiferum	Tannins	Anti-AIDS
Senecia jocobaea	Pyrrolizidine	Anti-tumor
Tabebuia avellandedae	Lapachol *(napthoquinones)*	Walker 256 Carcinosarcoma
Taxus baccata	Taxol (diterpene)	Anti-cancer
Tripterigium wilfordii	Tripdiolide Triptolide (diterpene triepoxide)	Anti-cancer
Quassia amara	Quassin (triterpenes)	Anti-tumour

An improvement in the product formation was felt necessary to accomplish the demand of mankind. *In vivo* grown plants has certain intrinsic limitations, such as, production of the bio-active compounds are seasonal in many cases and dependant on organ development, thus limiting production throughout the year. Plants are fixed in the environment and as such, they are often subjected to natural calamities and pathogen attack, which restrict the stability of product formation. The yield often varies in different ecological condition. To circumvent the above mentioned short falls and to improve the production of desirable compounds, non-conventional method of *in vitro* techniques are being employed.

Under uniform cultural condition, metabolites could be produced uniformly throughout the year with consistent quality, free from pests and diseases and without interference of environmental hazards. Using the established protocol, the product synthesis can be reproduced even in geographically distant places.

Tissue culture technique as it stands today is a mere empirical science, because, most of the experimental results can not yet be explained at the cellular or molecular level. Moreover, biological materials are widely diverse in their genetic make up, as such, they respond differently under the given situation. Therefore, for every plant species, standardisation of medium composition, explant types, temperature, photoperiod etc. are essential.

Various factors, such as, biological, chemical and physical, play a vital role in improving product synthesis has been illustrated in the following chapters, alongwith differentiation at the cellular and morphological level. Use of elicitors, precursors etc. to enhance the desirable product has also shown to hold promise. Bio-transformation technique has also been employed in conversion of intermediary compound produce abundantly in an organism to a desirable compound using an enzyme from microbial source.

Factors Affecting Medicinal Compound Production *In Vitro*

Nutrient Media

Success of plant tissue culture depends on the choice of nutrient medium. The tissues of most plant species can be grown on a completely defined medium. The media formulated by Murashige and Skoog (1962); White (1963); Schenk and Hildebrandt (1972) and B_5 of Gamborg *et al.* (1968) and their modifications are most widely used for culture of cell, tissues and organs of different plants yielding secondary metabolites.

In case of plant tissue culture, MS medium is mostly used for callus initiations and proliferation for the production of secondary metabolites, e.g. species of *Nicotiana*, *Solanum, Balanites, Phellodendron, Datura, Rauwolfia, Dioscorea* etc. Use of B_5, SH and White's media are also reported in other plant species. Generally, medium conditions which mostly support synthesis of secondary metabolites are those which limit rapid cell division and lead to a comparatively short exponential phase of growth.

Components of Nutrient Media

The chemical composition of any medium has primarily been selected based on the aim of work. Consequently, a method has to be formulated to produce a particular plant product. A compromise between two sets of optimum conditions, that is, those for accumulation of biomass and those for the biosynthesis and accumulation of secondary product, is best striven for.

West and Henshaw (1974) have shown that formation of catechol, tannins in sycamore suspension culture is enhanced by increasing the ratio of the carbon to nitrogen sources. Hahlbrock *et al.* (1974) suggested that exhaustion of nitrogenous nutrients may induce a shift in the cellular metabolism to reactions forming nitrogen-free products. For example, phenylpropanoid metabolism would continue increasing sugar of the medium and lowering nitrogenous compounds. Zenk *et al.*, (1975) observed that anthroquinone synthesis in *Morinda citrifolia* was not inhibited by raising the nitrate concentration. Maximal pigment formation occurred in the presence of 5 gm/L potassium phosphate. Carew and Krueger (1977) noted that indole compounds increased in culture of *Catharanthus roseus* when level of phosphate was raised in the medium.

The effect of inorganic macronutrients on cell growth and diosgenin production was studied in *Costus speciosus* cell culture. Schenk and Hildebrandt's (1972) basal medium was found to be suitable for culture of zygotic embryos. The medium was negatively or positively modified depending on the concentration of nitrate, ammonium and phosphate. Among these, phosphate had the most significant effect when decreased to half or quarter concentrations, approximately 60 and 50% increase of the diosgenin content respectively, was recorded. On the other hand, increase in nitrate also had positive effect on diosgenin production, whereas, increase in ammonium of phosphate did not affect the diosgenin productivity (Pal, unpublished).

Carbon Source

There are few reports on the relative effectiveness of various carbohydrates as carbon sources for secondary metabolite formation by cell suspension cultures (Matsumoto *et al.*, 1973). Zenk *et al.*, (1975) tested 14 carbohydrates at 2% (W/V) levels and found that sucrose gave the highest yield of anthraquinones in *M. citrifolia* cultures. Ikeda *et al.*, (1976) have reported that production of ubiquinone -10 is increased when less than the standard amount of sugar is added to the medium. Suther *et al.*, (1980) have shown clearly that sucrose is more effective than glucose, fructose and maltose for growth and disogenin production. Fujita *et al.*, (1981) have demonstrated that production of shikonin derivatives in *L. erythrorhizon* depended on the concentration of sucrose when it was less than 2%. But shikonin production was inhibited at > 5% sucrose with optimum concentration of sucrose was < 3%. Ajmalicine production could be enhanced in *Catharanthus roseus* by increasing 8% sucrose in the medium, although there is a decrease in biomass growth (Stafford, 1991). In *Chrysanthemum cinerariaefolium* the optimum concentration of sucrose was 3% which stimulates both growth and pyrethrin production (Dhar, 1987).

Growth Regulators

Plant growth regulators are effective inducers of secondary metabolism (Bohm, 1980). Similarly both quality and quantity of auxins initially present in the tissue or administered during the course of culture development have a marked effect on secondary metabolite production (Gamborg *et al.*, 1971).

Mizusaki *et al.* (1971) demonstrated that high levels of auxins inhibited the activity of putrescine-N-methyl transferase catalysing the N-methylation of putrescine, the key intermediate in the biosynthesis of nicotine. Konoshima *et al.* (1970) reported that kinetin inhibits tropane alkaloid synthesis in *Datura* tissue cultures. Shiio and Ohta (1973) along with Takahashi and Yamada (1973) reported that lower concentration of auxins viz. IAA, NAA and 2,4-D promote nicotine synthesis in tobaco cell cultures and higher concentrations of auxins inhibit nicotine synthesis in tobacco cell suspension cultures. Kurz and Constabel (1979) reported that 2,4-D lowers the pool of glutamate and aspartate (an important amino acid precursor of many alkaloids) which are present in the inhibited biosynthesis of shikonin in *Lithospermum* cultures (Tabata *et al.*, 1974). On the other hand, 2,4-D stimulates the production of compounds like diosgenin in cultures of *Dioscorea* (Kaul *et al.*, 1969) and L-DOPA in cultures of *Mucana* (Brain, 1976). 2,4-D combined with BAP at a lower concentration promotes synthesis of pyrethrin in *Chrysanthemum cinerariaefolium* (Sarkar and Pal, 1991; Dhar and Pal, 1993).

There are only a few reports of metabolite level being increased by raising auxin levels, e.g. maximum carotenoid and ubiquinone contents of carrot cell cultures induced by the comparatively high level of 2,4-D (10 mg/L) (Ikeda *et al.*, 1976). A perusal of literature revealed that 2,4-D is used at a concentration ranging from 0.1 to 5.0 mg/L, except in cultures of *Mucana pruriens,* where 25 mg/L 2,4-D is optimal for L-DOPA production (Brain, 1976).

In case of *M. citrifolia,* anthraquinone synthesis was completely inhibited when callus was maintained for long in the medium containing 2,4-D (Zenk *et al.,* 1975). However, 2,4-D did not affect the synthesis of emodine type anthraquinone in *Cassia tora* cultures (Tabata *et al.,* 1975). In cultures of *L. erythorhizon* BA enhanced shikonin production (Misawa, 1985). In *Coptis japonica* cultures, among the various combinations of cytokinins (BA, Kn) and auxins (2,4-D, IAA, IBA, NAA) tested, BA (0.22 mg/L or 0.2 mg/L) and NAA (18.6 mg/L) were most favourable for both cell growth and alkaloid production. In contrast, BA at a higher concentration (2.2 mg/L) caused a significant decrease in the contents of plamitive or jateorrhizine without affecting the production of berberine and coptisine (Fukui *et al.,* 1982). Samejima *et al.* (1982) had shown that in callus cells of *Cryptomeria japonica* higher concentration of NAA repressed the accumulation of flavonol compounds, whereas, 2,4-D is less effecttve. At all Kn levels flavonols accumulated abundantly while high concentrations of BA repressed the accumulation. GA increased flavonol accumulation in calli. The effect of contrasting levels of the BA on the accumulation of steroidal sapogenins in *Solanum aviculare* was observed by Mantell and Smith (1983). Highest accumulations were achieved in media containing 0.02 mg/L of BA.

It was evident from the published literature that low levels of auxins have stimulatory effects on *in vitro,* secondary metabolite synthesis and higher levels of auxins have inhibitory effects. Low levels of BA too have stimulatory effects on some secondary metabolite formation in tissue cultures.

Growth Kinetics

The time period required to synthesize secondary metabolite in culture reflects the time needed to initiate the expression of the gene(s), responsible for respective enzyme activity. Kinetic studies, on the rate of metabolite formation and cell growth are essential to obtain a basic understanding of secondary metabolite production in plant tissue culture. Little information is available till date on the relationship between age of particular cell and product synthesis. Tabata (1977) has classified the growth stage and production pattern of cell culture into three major types.

- In the first type metabolite synthesis proceeds almost parallel to the cell growth (nicotine, tropane alkaloids, anthraquinone, pyrethrin).
- Secondary metabolites produced during the stationary phase (polyphenols, shikonin, sesquiterpene lactones, serpentine etc.).
- The metabolite produced twice in a growth cycle.

When secondary metabolites are produced which paralleled with cell growth, then a part of the intermediates from primary pathway are diverted towards the synthesis of secondary product. In the second category, synthesis of secondary compounds starts after completion of primary metabolite synthesis, hence the pool of intermediates are available for secondary metabolite synthesis. While in the last type, synthesis starts when primary metabolism has not started or declined.

Light

Normally, cultures do not grow photoautotrophically (except few cases, Yamada *et al.,* 1978). The behaviour of cultures is influenced by photoperiodicity, light quality and intensity (Seibert and Kadkade, 1980). The activities of the enzymes involved in the biosynthesis of cinnamic acids, coumarins, lignins, flavones, flavonols, chalcones and anthocyanins are significantly influenced by light intensity and photoperiodicity (Hahlbrock *et al.,* 1980).

Light induced polyphenol biosynthesis is known from rose (Forrest, 1969) and tea (Davies, 1972) culture. Upon illumination, the tea callus culture produces several fold increase in catechin, epicatechin and leucoanthocyanin. Cultures of *Happlopappus gracilis* kept under darkness three days prior to light exposure accumulate more anthocyanin than did cultures kept in dark for one day prior to light exposure (Fritsch *et al.,* 1971; Stickland and Sunderland, 1972). It has been demonstrated that enzyme involved in flavone accumulation increased with UV light, thus, induces enhanced flavonoid production. Blue light inhibited shikonin (ubiquinone) production in *Lithospermum erythrorhizon* (Tabata *et al.,* 1974). The synthesis of terpenoids and few alkaloids also inhibited when exposed to blue light, whereas, production of these bioactive compounds enhanced *in vitro* under darkness and also when exposed to red light (Nagel and Reinhard, 1975).

In contrast to trigerring effects of light on bioactive compound production *in vitro,* a number of reports indicate that light can have inhibitory effects, e.g. alkaloid production in *Scopolia parviflora* (Tabata *et al.*, 1972), suppressed L-DOPA accumulation in static cultures of *Mucuna pruriens* (Ohta and Yatazawa, 1978) and 70% suppression of nicotine production in *Nicotiana tabacum* callus by high light intensities (Weeks, 1976).

Genotypes

Like *in vivo* system, *in vitro* study also revealed that there are qualitative and quantitative differences in secondary metabolite production with the genotypes of plant under investigation. Interestingly, it is not always true that high yielding plants produce more metabolite under cultural condition. It has been demonstrated that serpentine concentration of cell culture established from a high yielding *Catharanthus* genotypes was not necessarily higher than that of the cell culture raised from a plant with low serpentine content. This is not surprising because the genes controlling the quantitative feature is remarkably influenced by the cultural environment. In *Peganum* a poor co-relationship exists between the alkaloid content of callus of cell suspension culture lines and parent plants (Sasse *et al.,* 1982).

Investigations have revealed that *Ammi visnaga* and *Ruta graveolens* produced characteristic secondary metabolites, in amounts similar to those in the plant from which they were derived. Kinnersly and Dougall (1980) reported that callus cultures of two low-alkaloid containing lines, of *N. tabacum* L. yielded considerably lower nicotine contents than cultures from the respective high-alkaloid yielding cultivars which were isogenic except for two loci for alkaloid accumulation. Thus, there was a strong correlation between the nicotine content of callus cultures and the plants from which they were derived. Zieg *et al.* (1983) reported that few high pyrethrin yielding strains were derived from high yielding parent plants of *Chrysanthemum cinerariaefolium.*

Role of Morphological and Cellular Differentiation in Product Synthesis

A major disdvantage of plant cell culture compared with the whole plant with respect to secondary metabolism is that the biosynthetic potential of the plant in question is very often not expressed in culture. Furthermore, it is desirable to use morphologically undifferentiated cells for the production of a large amount of desirable metabolites, as a potential alternative to whole plant source. At present total synthesis of complex biochemicals such as medicinal compounds from simple, primary precursors supplied in the medium is rarely achieved at a commercially significant level in suspension culture. On the other hand, there are many examples which show high productivity of undifferentiated cells than that of the intact plants, some of which are tabulated in Table 2.

TABLE 2: HIGH YIELDING CELL LINES WHICH ACCUMULATE BIO-ACTIVE COMPOUNDS AT LEVEL HIGHER THAN THE PARENT PLANT

Plant Species	Product	Yield (%dry weight)
Ammi visnaga	Visnagin	0.31
Berberis parvifolia cell line 177	*Jatrorrhizine*	
	Palmatine	
	Berberine	5.83
Cassia tora	Anthraquinones	6.0
Catharanthus roseus	Vindoline	
	Catharanthine	0.13
	Serpentine	0.8
	Ajmalicin	1.0
Cephaelis ipecacuanha	Emetine	
	Cephaelin	2.2
Coffea arabica	Caffeine	1.6
Coleus blumei	Rosmarinic acid	15.0
Coptis japonica	Berberine	11.4
Dioscorea deltoidea	Diosgenin	1.5
Ephedra sp.	Pseudoephedrine	2.25
Fedia cornucopiae	Valtrate	1.38
Lithospermum erythrorhizon	Shikonin	14.0
Macleaya microcarpa	Protopine	0.4
Morinda citrifolia	Anthraquinones	18.0
Nicotiana rustica	Nicotine	0.29
Nicotiana tabacum	Nicotine	3.4
	Glutathione	1.0
Panax ginseng	Ginsengoside	27.0
Papaver somniferum	Papavarine	5.6

Therefore, studies have been initiated using differentiated cells which have roots, shoots or other organs, without knowing clearly that to what extent secondary metabolites depends on the development of specific structures/organs, or, whether these two processes are genetically and/ physiologically linked.

Dhar and Pal (1988) have demonstrated that pyrethrin is being synthesized more in *Chrysanthemum cinerariaefolium* cultures with shoots than the roots and its content is even more lower in undifferentiated callus culture. Similar observations were also reported in cultures of *Datura metaloides,* where tropane alkaloid production increased in the root and shoot forming cultures than undifferentiated ones.

Thus, it seems that in many cases morphological differentiation may be necessary to obtain higher yields of secondary metabolites. However, this is not always desirable in large scale cultivation because, generally the culture period for differentiated tissue is long than that of undifferentiated cells. It is obvious that a shorter culture period is required to avoid microbial contamination and to lower the production cost.

It is known that essential oils and flavour compounds in plants accumulate in morphologically specialised structures, such as glands in mint, oil ducts in celery, swollen leaf bases in onion, secretory ducts, latex components found in laticifers. In *C. cinerariaefolium,* pyrethrins are localized in the oil glands and ducts present on the achenes (Chandler, 1951) and on leaves (Zito *et al.,* 1983). We have shown for the first time through scanning electron microscopy that the oil glands were also distributed all over the surface of the cells of callus mass derived from leaf and petiole explants (Pal and Dhar, 1984). These oil glands are the site of pyrethrin biosynthesis. This finding further suggests that there exists a correlation between the number of oil glands and amount of pyrethrin. The higher the number of glands, more the pyrethrin content.

Both isolated roots and root callus of *Atropa belladonna* contain measurable quantity of atropine while stem and leaf callus, devoid of macroscopic roots, do not possess the alkaloid. Rotenoids were detected in callus tissue culture of *Derris elliptica* with differentiated roots. *In vivo* rotanoids are being present in the roots of *D.elliptica*.

Role of Precursors

Attempts have been made to increase the contents of secondary metabolites in plant cell cultures by feeding related precursors and intermediates. Cell suspension cultures of *Coleus blumei* have been reported to produce about 100% more rosmarinic acid than the control, when supplied with 500 mg α-phenylalanine/L (Zenk *et al.,* 1977). Feeding experiments of radioactively labeled phenylalanine and tyrosin to suspension cultures of *C.blumei* showed that phenylalanine was mainly incorporated into the caffeic acid part of the molecule, and tyrosine gave rise to the 3,4-dihydroxyphenyllactic acid moity (Peterson, 1994). Anthraquinone production in cell suspension cultures of *Gallium mollugo* was increased by supplying o-succinyl benzoic acid (Wilson, 1980). Bauch and Leistner (1978) were of the opinion that non-production of secondary metabolites (viz. iridoids) in the *Gallium* cell suspension culture is not due to a lack of appropriate biosynthetic precursors, but due to the absence or presence of metabolites (viz. anthraquinones and naphthols) reflects the balance between their biosynthesis and degradation. Cholesterol, a steroidal precursor, when fed at 100 mg/L concentration to suspension cultures of *Dioscorea deltoidea,* approximately 100% enhancement of diosgenin content was recorded by Chowdhuri and Chaturvedi (1979).

Precursors of a particular product often prove more suitable as stimulators under some specific cultural conditions. For instance, feeding tobacco cultures with the nicotine precursors putrescine and nicotinic acid did not led to any measurable increase in nicotine biosynthesis. But, when cultures were fed with ornithine, the nicotine levels increase (Ohta *et al.*, 1978).

Role of Elicitors

Elicitors are a special group of triggering factors, which enhance the production of bioactive compounds. Mostly, these compounds are isolated from microorganisms and are responsible for stimulating particular facets of secondary plant metabolism.

Exposure of cell culture to the action of some elicitors, such as, fungal homogenates and fractions is well documented. Using homogenates obtained from the pathogenic *Botrytis,* changes were produced in alkaloid production, depending on the concentration and time of introduction of elicitor into the medium containing *Berberis parviflora* cell suspension culture (Deliu *et al.,* 1994). The productivity of cultures significantly increased from 0.5 g protoberberine alkaloids to 0.8 g/L, in which 3 ml of *Botrytis* homogenate was added to 100 ml of culture medium, 8 hrs before harvesting the *B. parviflora* cells. In this treatment, release of alkaloids into the medium was mainly stimulated without loss of viability in *Berberis* cells.

Polysaccharide elicitors Pms, after obtained from *Phytophthora megasperma* var. *sojae*, when added to soyabean cell suspension cultures growing in the dark induced glyceolin formation after 12 hours and accumulates thereafter (Ebel *et al.*, 1978). Prior to this development, a rapid increase of PAL activity starts, reaching to a maximum level 20 hrs after the addition of Pms. Comparative density—labelling technique revealed that such increases in PAL activity associated with elicited responses, are partly due to *de novo* synthesis of this enzyme (Lamb and Dixon, 1978).

Application of Bio-Transformation Technique

Bio-transformation is a process through which the functional groups of organic compounds are altered by living cells (Stepan-Sarkissian, 1991). The technique of bio-transformation was initially applied to the microbes due to a number of inherent advantages. Attention in recent years has been focused on the possibility of production of bioactive compounds at a commercially significant level by inducing them to perform special bio-transformation reactions on organic compound added to the medium and not necessarily present in the species under investigation. This way a substance of lower value is transformed to one of higher value and is also available in substantial amounts.

In cases, where the final desirable product of a given pathway is not available from cells cultivated *in vitro,* due to non-expression of one or more enzymes in the intermediate stages of the pathway, the supply of immediate or near precursors and their conversion into the desired product can be possible. Alternatively, the bio-transformation of synthetic substances, analogues, or secondary metabolites from other plant species can result in the production of novel compound(s) hitherto unknown in nature, which have new and unique chemical and pharmacological properties.

Bio-transformation technique has also been employed in suspension culture for conversion of intermediary compound produced abundantly to a desirable compound using an enzyme either from microbial or from plant source. In nature, desoxypodophyllotoxin is more frequently encountered than podophyllotoxin and therefore, used as an alternative raw material for synthesis of anti-tumor lignan, viz. Podophyllotoxone and picropodophyllotoxone in *Forsythia x intermedia* cell suspension cultures, presumably via benzylic hydroxylation and then oxidation (Broomhead and Dewick, 1991).

Biotransformation of β–methyl digitoxin to β–methyl digoxin using *Digitalis lanata* cells was demonstrated by Reinhard and Alfermann (1980). Digoxin has a big market as a cardiac glycoside. Kurz *et al.,* (1985) found that 3'-4' dehydrovinblastine was converted to leurosine and catharanthine which also accumulate in *C. roseus* cells. The biotransformation of thebaine to neopine by *Papaver somniferum cv Marianne* cell suspension cultures was also demonstrated.

Application of Cell Immobilization Technique

The property of the cells to adhere to many surfaces in nature is known for quiet sometime. Traditional use of immobilized *Acetobactor* cells for the production of vinegar is well known.

Immobilization has been defined as a technique which confers a catalytically active enzyme or cell within a reactor system and prevents its entry into the mobile phase which carries the substrate and product (Scragg, 1991). Immobilization of plant cells was first reported by Brodelius and colleagues in 1979. They have shown that entrapment of *Catharanthus roseus* and *Daucas carota* cells in alginate (or other polysaccharides) matrix enhanced product synthesis. The mechanism involved in the process is physically restraining the culture cell within a solid matrix such as alginate, polyurethane foam and wood pulp etc.

Secondary metabolites are generally associated with slow growing compact cells. Through immobilisation of cultured cells, the cell aggregation process can be encouraged and the physical contact often induces cyto-differentiation, which in turn promotes increased metabolite formation.

The most commonly used matrix for cell entrapment is calcium alginate beads. Polyurethane foam has only rarely been used as immobilising agent (Corchete and Yeoman, 1989). Washed and dried wood-pulps (0.5 gm) was added per 50 ml of medium in a conical flask and sterilised, prior to subculture of the cells. The suspension culture of *Papaver somniferum* was grown for 28 d, harvested, dried and assayed. Both polyurethane foam and wood pulp enhanced alkaloid production without significantly increasing biomass yield. Wood pulp proved to be superior in all experiments, resulting in almost a doubling of dihydrosan-guinarine yields (mg/L), when compared to controls. The increased alkaloid yields was explained by same chemical interaction between the cellulosic cell walls of the wood pulp and/or enzyme systems of the cultured cells.

Morinda citrifolia cells were immobilised by entrapment in alginate matrix to produce anthraquinone (Brodelius and Nilsson, 1983). It was found that the immobilised cells had an increased synthesis of anthraquinones as compared to freely suspended cells under the same conditions. The researcher has also demonstrated that the synthesis of indole alkaloids (100%) yield from tryptamine and secologanine can be activated using immobilised *C. roseus* cells. Although indole alkaloids are normally stored in the vacuoles of the cell in addition of a trace amount of chloroform in the aqueous phase, altered cell permeability and consequently the products were excreted extracellularly. In a comparison with several types of polysaccharides as matrices, it was recognised that alginate entrapped *C.roseus* had the highest productivity of ajmalicine, a vasodilator.

The use of immobilized plant cells (vide pg. 112) for bio-transformations has covered a wide range of reactions including hydroxylations, methylations and glycosylations. The best example is bio-transformation from β-methyldigitoxin to β-methyldigoxin using alginate entrapped *Digitalis lanata* (Alfermann *et al.,* 1983). Biotransformation occurred at a constant rate for at least 150 d after a lag phase of about 20 d.

Advantages of Using Immobilised Cells

- Expensive biocatalysts can be reused.
- Allows continuous operation
- Aggregation of cells can be readily controlled.
- Products are easily separated from the catalysts.
- Reduction in genetic unstability.

Following Conditions are Pre-Requisites for a Successful Immobilisation Process

- The product of interest should not be associated with growth of the tissue. Therefore, the compound enlisted in group of Tabata's classification (vide pg. 108 para 4) are not suitable for such work.
- Slow growing cultures are preferred to prevent disintegration of the biocatalyst matrix and eventually disruption of the entire system.
- Immobilized cells should remain viable for longer period and should have uniform, sustainable yield of bioactive compound production.

Application of. Transgenic Technique

One recent advance in transgenic technology towards over production of medicinal compounds is application of transgenic organ culture such as shooty teratomas or hairy roots.

An efficient means of producing bioactive compounds that are normally biosynthesised in leaves of plants, is practised by incorporating T-DNA of Ti plasmid of *Agrobacterium tumefaciens.* Tumour forming Ti plasmid produced shooty teratomas under *in vitro* condition resulted in enhanced biomass production and *de novo* biosynthesis of some specific bioactive compounds.

Plant tumours have greater metabolic activity than normal tissues, and may therefore be useful as sources of biologically active compounds. For example, tumour of sugar beet (*Beta vulgaris)* and tomato contain more ascorbic acid and glutathione than normal tissue. Crown gall tumours appear to reflect the physiological state of the host plant from which they developed. The tumour grows indefinitely on very simple, hormone free medium with only a few organic supplements, so that cultivation costs are lower than normal tissues. Crown gall is easily induced on a large number of gymnosperms and dicotyledonous plants by wounding and inoculation with *A. tumefaciens.* Plants susceptible to crown gall include Juniperus (monoterpene, diterpene, sesquiterpene producer), *Apocynum cannabium* (cardioglycoside), *Catharanthus roseus* (ajmalicine, catharanthin, vincristin, vinblastin), *Nerium oleander* and *Digitalis purpurea* (cardioglycoside), etc. (De Cleene, 1984).

On the other hand, the hairy roots can be induced following genetic transformation of plant tissues with the Ri plasmid of the soil bacterium. *Agrobacterium rhizogenes* have been proved to be an efficient means of producing bioactive compounds that are normally biosynthesised in roots of *in vivo* grown plants. These are genetically stable culture, capable of producing the compounds characteristic of normal plant roots. On a laboratory scale, transformed root cultures are grown in a liquid medium containing sucrose and mineral salts including a nitrogen source devoid of any plant growth substances (auxins and cytokinins). This is because of the fact that transformation process involved transfer of genes from the *Agrobacterium,* which alter the metabolism of auxins and cytokinins and promotes root production. The cultures can be grown either in a 1 to 5 L. bioreactor or in pilot scale bioreactor of 500 L. as in AFRC Institute of Food Research (Norwich).

Hairy root cultures of *Lithospermum erythrorhizon* were established by transformation of *in vitro* grown shoots with *A. rhizogenes* 15834. This root cultures produced a large amount of red pigments. Using this method an air-lift fermenter system was established, equipped with a XAD-2 (as adsorbant) column, which continuously produced ca. 5 mg/day of shikonin during a period of more than 220 days (Shimomura *et al.,* 1991).

Hairy roots of *Panax ginseng* transformed by *Agrobacterium rhizogenes,* synthesized 2.4 times more ginsenosides, than that of native roots and about 2 times than that of cultured roots, on dry weight basis (Yoshikawa and Furuya, 1987).

Hairy root cultures induced by *A. rhizogenes* for the enhanced production of following bioactive compounds were also reported in the last decade: alkaloids from *Hyoscyamus muticus, Aconitum heterophyllum, Atropa belladona, Datura quercifolia, Rouwolfia serpentina*; isoprenylated flavonoids by *Glycyrrhiza glabra,* ginkgolide from *Ginkgo biloba*; catharanthine, ajmalicine and vindoline from *Catharanthus roseus,* steroidal alkaloids from *Solanum aviculare, S. mauritianum;* anthraquinones from *Rubia peregrina, R. tinctorum;* with anolide from *Withania somnifera;* tanin from *Geranium thunbergii;* phytoecdysteroids from *Ajuga reptans var. atropurpurea*; cardenolides from *Digitalis lanata;* amarogentin, amaroswerin and xanthones from *Swertia japonica;* valepotriates from *Valerina wallichii, V. officinalis* var. *sambucifolia;* antimalarial compound artemisinin from *Artemisia annua* etc.

Application of Recombinant DNA Technology

Research on medicinal plants using the tools of molecular biology and recombinant DNA technology has started only since the last decade. Targeting of specific gene(s), the expression of which will lead to alteration of secondary metabolism, has been made in a number of medicinal plants. Development of transgenic plant with high level of secondary metabolic product-accumulation by various means such as over expression, antisense or co-suppression technology has opened up new vistas in the field of research on medicinal plants.

Understanding of the molecular mechanism for regulation of the secondary pathway and gene expression of the rate-limiting step of the reaction, in particular, are the prerequisites for genetic manipulation. Moreover, the genes responsible for plant alkaloid production can be functionally expressed in microorganisms to produce either single biotransformation steps or short biosynthetic pathways.

Now, the question comes how these particular gene(s) can be tailored to produce important pharmaceutical alkaloids. This can be done by introducing side pathways in a known plant biosynthetic pathway; eliminating undesirable side pathways or even accumulating biosynthetic intermediates depending on the case. For example, the enzyme responsible for the conversion of desired alkaloid at the terminal end of a known pathway (which is otherwise difficult to achieve by chemical synthesis) can be regulated heterologously in microorganism, so that the organism can produce the desired compound when the substrate or precursor is supplemented to the medium. Even a relatively short biosynthetic pathway consisting of three to four enzymatic steps can be successfully expressed in a microorganism resulting in the overproduction of the desired compound (Kutchan, 1995). However, biosynthetic pathways of most of the alkaloids involve twenty to thirty intermediate enzymatic steps, engineering of such long pathways in totality in a single microorganism is presently beyond the technologies available to the molecular biologists. This is because it is known till date that the genes encoding the biosynthetic enzymes are neither clustered nor co-ordinately controlled by one single operon. However, current technology is sufficiently equipped to alter the pathway in the plant cell and produce the desired alkaloid either in culture or in the field by introduction of a transgene (from another plant or a microorganism) or by interrupting an undesirable catabolic reaction by successfully blocking a side pathway.

Based on these principles, transfer and expression of engineered model genes into medicinal plants have already been demonstrated for production of metabolites in *Beta vulgaris* (Hamil *et al.*, 1987), *Digitalis purpurea* (Saito *et al.*, 1990) and *Atropa belladona* (Yamazaki, 1991).

Acknowledgements

Authors are indebted to the Director, Bose Institute for providing all facilities required during the preparation of this article. They are thankful to Dr. Amitabh Roy for critically reading the manuscript.

References

Alfermann, A. W., Bergmann, W., Figur, C., Helmbold, U. and Schwantag, D.,1983. Biotransformation of β-methyl digitoxin to β-methyldigoxin by cell cultures of *Digitalis lanata*, *Semin. Ser. Soc. Exp. Biol.* **18**: 67-74.

Bauch, H. J.and Leistner, E., 1987. Aromatic metabolites in cell suspension cultures of *Gallium mollugo* L. *Planta Medica* **33** (2) : 105 -133.

Bohm, H., 1980. The formation of secondary metabolism in plant tissue and cell cultures. *Intl. Rev. Cytol., suppl.* **11 B:** 183-208.

Brain, K. R., 1976. Accumulation of L-DOPA in cultures of *Mucana pruriens. Plant Sci. Lett.,* **7**: 157-158.

Brodelius, P., Deus, B., Mosbach, K. and Zenk, M. H., 1979. Immoblized plant cells for the production and transformation of natural products. *FEBS Letters,* **103**: 93-97.

Brodelius, P. and Nilsson, K., 1980. Entrapment of plant cells in different matrices. A comparative study. *FEBS Letters* **122**: 312-315.

Broomhead, A. J. and Dewick, P. M., 1991. Biotransformation of *Podophyllum* ligans in cell suspension cultures of *Forsythia intermedia. Phytochemistry,* **30**: 1511-1517.

Carew, D. P. and Krueger, R. J., 1977. *Catharanthus roseus* tissue culture: the effect of medium modification on growth and alkaloid production. *Lloydia,* **40 (4)**: 326-336.

Chandler, S. E., 1951. Botanical aspects of pyrethrum I. *Pyrethrum Post,* **2**: 1-19.

Chowdhury, A. R. and Chaturvedi, H. C., 1979. Cholesterol and biosynthesis of diosgenin by tuber callus of *Dioscorea deltoidea. Current Sci.,* **49** : 237-238.

Corchete, P. and Yeoman, M. M., 1989. Biotransformation of (-) -codeinone to (-) -codeine by *Papaver somniferum* cells immobilised in reticulate polyurethane foam. *Plant Cell Rep.* **8(2)** : 128-131.

Davies, M. E., 1972. Polyphenol synthesis in cell suspension cultures of Paul's Scarlet Rose. *Planta* **104**: 50-60.

De Cleen, M., 1984. The use of *in vitro* plant tumours for pharmaceutical purposes. *Plant Tissue Cell Cult. Appl. Crop Impr. Meet,* 577-578.

Deliu, C., Tamas, M., Dobrota, C. and Munteanu-Deliu, C., 1992. Kinetics of cellular growth and diosgenin biosynthesis in *Dioscorea caucasica* cell suspensions cultured in batch system. *Plant Sci.,* **85(1)** : 99-105.

Deliu, C., Tamas, M. and Ghiran, D., 1994. *Berberis* species: *In vitro* culture and the production of protoberbarine and other alkaloids. *In*: *Biotechnology in Agriculture and Forestry,* Vol. 28, *Medicinal and Aromatic Plants VII* (Ed. Y. P. S. Bajaj); 56-71, Springer Verlag, Berlin.

Dhar, K., 1987. Ph D., Thesis entitled *"Tissue Culture of Chrysanthemum cinerariaefolium Vis. with special reference to pyrethrin synthesis in callus cultures".* Calcutta University, Calcutta, India.

Dhar, K and Pal, A., 1988. Pyrethrin content in unorganised and organized callus cultures of *Chrysanthemum cinerariaefolium* clone HSL 801. *Trans. Bose Res.* **51** : 57-63.

Dhar, K. and Pal, A., 1993. Factors influencing efficient pyrethrin production in undifferentiated cultures of *Chrysanthemum cinerariaefolium* Vis. *Fitoterapia* **LXIV** : 336-340.

Ebel. J., Ayers, A. R. and Albersheim, P., 1976. Host-pathogen interactions : XII : Response of suspension-cultured soybean cells to the elicitor isolated from *Phytophthora megasperma var. sojae,* a fungal pathogen of soybeans. *Plant Physiology,* **57** : 775-779.

Forrest, G. I., 1969. Studies on the polyphenol metabolism of tissue cultures derived from the tea plant *(Camellia sinensis* L*). Biochemical Journal,* **113** : 765-772.

Fritsch, H., Hahlbrock, K. and Grisebach, H., 1971. Biosynthese von Cyanidin in Zellsuspensionkulturen von *Haplopappus gracilis. Zeitschrift fur Naturforschung,* **266** : 581.

Fujita, Y., Hara, Y., Suga, C. and Morimoto, T., 1981. Production of shikonin derivatives by cell suspension cultures of *Lithospermum erythrorhizon. Plant Cell Rep.* **1** : 61-63.

Fukui, H., Nakagawa, K., Tsuda, S. and Tabata, M., 1982. Production of Isoquinoline alkaloids by cell suspension cultures of *Coptis japonica. In: Proc. 5th Intl. Congr. Plant Tissue and Cell Culture.* (ed. A. Fujiwara), Japan, pp. 353-354.

Gamborg, O. L., Constabel, F., La Rue, T. A. G., Miller, R. A. and Steek, W., 1971. Influence of hormones on secondary metabolite formation in plant cell cultures. In *Les. Culture de Tissue de Plantes, France,* pp. 335-344.

Gamborg, O., Miller, R. and Ojima, K., 1968. Nutrient requirements of suspension cultures of soybean root cells. *Exp. Cell Res.,* **50** : 151-158.

Hahlbrock, K., Ebel, J., Oaks. A., Auden, J. and Liersch, M., 1974. Determination of specifc growth stages of plant cell suspension cultures by monitoring conductivity changes of the medium (Soyabeans). *Planta.* **118 (1)** : 75-84.

Hahlbrock, K., Schroder, J. and Vieregge, J., 1980. Enzyme regulation in parsley and soyabean cell cultures. *In*: *Advances in Biochemical Engineering,* Vol. **18**, *Plant Cell Cultures II* (Ed. A. Feichter), pp. 39-60, Spinger-Verlag.

Hamill, J. D., Prescott, A. and Martin, C., 1987. An assessment of the efficiency of cotransformation of the T-DNA of *Agrobacterium rhizogenes* and the T-DNA of disarmed binary vectors derived from *Agrobacterium tumefaciens. Plant Mol. Biol.* **9** : 573-576.

Ikeda, T., Matsumoto, T. and Noguchi, M., 1976. Formation of ubiquinone by tobacco plant cells in suspension culture. *Phytochemistry,* **15** : 568-569.

Kaul, B., Stohs, S. J. and Staba, E. J., 1969. *Dioscorea* tissue cultures. III. Influence of various factors on diosgenin production by *Dioscorea deltoidea* callus and suspension cultures. *Lloydia*, **32** : 347-359.

Kinnersley, A. M. and Dougall, D. K., 1980. Correlation between the nicotine content of tobacco plants and callus cultures. *Planta*, **149** : 205-206.

Konoshima, M., Tabata, M., Hiraoka, N. and Miyake, H., 1990. Growth and alkaloid production of *Datura* tissue culture. *Yakugaku Zasshi*, **90** : 370-377.

Kurz, W. G. W. and Constabel, F., 1979. Plant cell cultures, a potential source of pharmaceuticals. *Advances in Applied Microbiology*, **25** : 209-240.

Kurz, W. G. W., Chatson, K. B. and Constabel, F., 1985. Biosynthesis and accumulation of indole alkaloids in cell suspension cultures of *Catharanthus roseus* cultivars. *In : Primary and Secondary Metabolism of Plant Cell Cultures*. (Eds. K.H. Newmann, W. Barz, and E. Reinhard), pp. 143-153.

Kutchan, T. M., 1995. Alkaloid Biosynthesis—the basis for metabolic engineering of medicinal plants. *The Plant Cell*, **7** : 1059-1070.

Lamb. C. J. and Dixon, R. A., 1978. Stimulation of *de novo* synthesis of L-phenylalanine ammonia-lyase during induction of phytoalexin biosynthesis in cell suspension cultures of *Phaseolus vulgaris*. *FEBS Letters*, **94** : 277-280.

Mantell, S. M. and Smith, H., 1983. Cultural factors that influence secondary metabolite accumulations in plant cell and tissue cultures. *In*: *Plant Biotechnology* (Eds. S. M. Mantell and H. Smith) Cambridge University Press, London, pp. 75-108.

Matsumoto, T., Nishida, K., Noguchi, M. and Tamaki, E., 1973. Some factors affecting the anthocyanin formation by *Populus* cells in suspension cultures, *Agri. Biol. Chem.*, **37** : 561-567.

Misawa, M., 1985. Production of useful plant metabolites *In* : *Proc. 36th Ann. Meeting of Tissue Cultures Association*, Japan, pp. 59-87.

Mizusaki, S., Tanabe, Y., Noguchi, M. and Tamaki, E., 1971. p-coumaroyl-putrescine, Caffeoylputrescine and feruloylputrescine from callus tissue of *Nicotiana tabacum*. *Physiol. Plant*. **15** : 473-497.

Ohta, S., and Yatazawa, M., 1978. Effect of light on nicotine production in tobacco tissue culture. *Agricultural and Biological Chemistry*, **42** : 873-877.

Ohta, S., Matsui, O. and Yatazawa, M., 1978. Culture conditions for nicotine production in tobacco tissue culture. *Agricultural and Biological Chemistry*, **42** : 1245-1251.

Nagel, M. and Reinhard, E., 1975. Das athersche ol der callus kulturen von *Ruta graveolens*. II. *Physiologie zur bildung des atherischen oles. Planta med.*, **27**: 264-274.

Pal, A. and Dhar, K., 1984. Application of tissue culture for pyrethrin production *In: Proc Symp. on Applied Biotechnology of Medicinal, Aromatic and Timber Yielding Plants*, (Ed. P. C. Datta), NSABMATP, Calcutta University, Calcutta, pp. 70-80.

Peterson, M., 1994. *Coleus* spp : *In vitro* culture and the production of forskolin and rosmarinic acid. *In* : *Biotechnology in Agriculture and Forestry* **26** : *Medicinal and Aromatic Plants* VI, Ed. Y. P. S. Bajaj, Springer-Verlag, New York, pp. 69-92.

Reinhard, E. and Alfermann, A. W., 1980. Biotransformation by plant cell cultures. *Adv. Biochem. Eng.* **16** : 49-83.

Saito, K., Yamazaki, M., Kaneko, H., Nojii, M., Imai, Y. and Murakoshi, I., 1990. Transgenic Medicinal plants : Introduction of genes for drug metabolizing enzymes, *J. Cell Biochem.*, Suppl., **14 E** : 357.

Samejima, M., Yamaguchi, T., Fukuzumi, T. and Yoshimoto, T., 1982. Effects of phytohormones on accumulation of flavonols in callus culture. *In: Proc. 5th Intl. Congr. Plant Tissue and Cell Culture* (ed. A. Fujiwara), Japan pp. 353-354.

Sarkar, K. and Pal, A., 1991. Factors affecting stability in pyrethrin production in cultures of *Chrysanthemum cinerariaefolium* Vis., *Acta Botanica Indica*, **19**: 248-251.

Sasse, F., Knobloch, K. H. and Berlin, J., 1982. Induction of secondary metabolism in cell suspension cultures of *Catharanthus roseus, Nicotiana tabacum* and *Peganum harmala*. *In* : *Proc. 5th Intl. Congr. Plant Tissue and Cell Culture* (ed. A. Fujiwara), Japan, pp. 343-344.

Schenk, R. U. and Hidebrandt, A. C.,1972. Medium and techniques for induction and growth of monocotyledonous and dicotyledonous plant cell cultures, *Can. J. Bot.* **50** : 199-204.

Scragg, A., 1991. The immobilization of plant cells. *In* : *Plant Cell and Tissue Culture*, (Eds A. Stafford and G. Warrens), The Institute of Biology, U. K., 205-219.

Seibert, M. and Kadkade, P. G., 1980. Environmental factors: A Light. *In* : *Plant Tissue Culture as a Source of Biochemicals*. (Ed. E.J. Staba), Boca Raton, Florida: CRC Press, pp. 123-142.

Shiio, I. and Ohta, S., 1973. Nicotine production by tobacco callus tissues and effect of plant growth regulators. *Agr. Biol. Chem.*, **37** : 1857-1864.

Shimomura, K., Sudo, H. Saga, H. and Kamada, H., 1991. Shikonin production and secretion by hairy root cultures of *Lithospermum erythrorhizon*. *Plant Celll Rep.*, **10** : 282–285.

Stafford, A., 1991. Natural products and metabolites from plants and plant Tissue cultures. *In* : *Plant Cell and Tissue Culture,* (Eds A. Stafford and G. Warrens), The Institute of Biology, U. K., 124-162.

Stepan- Sarkissian, G., 1991. Biotransformation by plant cell culture. *In* : *Plant Cell and Tissue Culture*, (Eds A. Stafford and G. Warrens), The Institute of Biology, U. K., 163- 204.

Stickland, R. G. and Sunderland, N., 1972. Production of anthocyanins, flavonols and chlorogenic acids by cultured callus tissues of *Haplopappus gracilis. Annals of Botany,* **36 :** 443-457.

Suther, K. C., Ramawat, K. G. and Arya, H. C., 1980. Effects of sugars on diosgenin production in *Balanites aegyptica* cultures. *In* : *Proc. Plant Tissue Culture, Genetic Manipulation and Somatic Hybridization of Plant Cells* (Eds. P. S. Rao, M. R. Heble, and M. S. Chadha), BARC, Trombay, India, pp. 220-226.

Tabata, M., 1977. Recent advances in the production of medicinal substances by plant cell culture. *In* :*Plant Tissue Culture and its Biotechnological Application* (Eds. W. Barz, E. Reinhard and M. H. Zenk), Springer-Verlag, Germany, pp. 2-16.

Tabata, M., Yamamoto, H., Hiraoka, N. and Konoshima, M., 1972. Organization and alkaloid production in tissue cultures of *Scopolia parviflora. Phytochemistry.* **11,** 949-955.

Tabata, M., Mizukami, H., Hiraoka, N., Marumoto, Y. and Konoshima, M., 1974. Pigment formation in callus cultures of *Lithospermum erythrorhizon. Phytochem.*, **13 :** 927-932.

Tabata, M., Hiraoka, N., Ikenoue, M., Sano, Y. and Konoshima, M., 1975. The production of anthraquinones in callus cultures of *Cassia tora. Lloydia,* **38**: 131-134.

Takahsi, M. and Yamada, Y., 1973. Regulation of niotine production by auxins in tobacco cultured cells *in vitro. Agr. Biol. Chem.*, **37**: 1755-1757.

Weeks, W. W., 1976. *Physiology of alkaloids in germinating seed of Nicotiana tabacum. PhDThesis.* University of Kentucky, Lexington, USA.

West, C. R. J. and Henshaw, G. G., 1974. Phenolic metabolism in suspension cultures of *Acer psedo-platanus* L. *In*: *Abst. 3rd Intel. Congr. Pl. Tissue Cell Culture,* Leicester, U. K. No. 256.

Wilson, G., 1980. Continuous culture of plant cells using the chemostate principle. *In* : *Advances in Biochemical Engineering,* vol. 16. *Plant Cell Cultures,* Ed. A. Fiechter, pp. 1-25, Springer Verlag, Berlin.

White, P. R., 1963. *The Cultivation of Animal and Plant Cells (2nd edition),* Ronald Press. New York.

Yamada, Y., Sato, F. and Hagimori, M., 1978. Photoautotrophism in green culture cells. *In*: *Frontiers of Plant Tissue Culture* (Ed. T. A. Thorpe), University of Calgary, Calgary, Canada, pp. 453-462.

Yamazaki, M., 1991. *Transfer and expression of foreign genes into medicinal plants and production of secondary metabolites*, Ph. D. Thesis (in Japanese), Chiba University, Chiba 260, Japan.

Yoshikawa, T. and Furuya, T., 1987. Saponin production by cultures of *Panax ginseng* transformed with *Agrobacterium rhizogenes. Plant Cell Rep.*, **6** : 449-453.

Zieg, O. R. G., Zito, S. W. and Staba, E. J., 1983. Selection of high pyrethrin producing cultures. *Planta Med.*, **98** : 88-91.

Zenk, M. H., El-shagi, H. and Schulte, U., 1975. Anthraquinone production by cell suspension cultures of *Morinda citrifolia. Planta Med.* (Supplement) : 79-101.

Zenk, M. H., El-Shagi, H. and Schulte, U., 1977. Anthraquinone production by cell suspension cultures of, *Coleus blumei. Naturwissenschaften,* **64** : 585-586.

Zito, S. W., Zieg, R. G. and Staba E. J., 1983. Distribution of pyrethrin in oil glands and leaf tissues of *Chrysanthemum cinerariaefolium. Planta Med.*, **47** : 205-207.

Recent Trends in Spices and Medicinal Plants Research pp. 122-124, 2000
Ed. Amit Krishna De
Associated Publishing Co., New Delhi

2.9

INFLUENCE OF GROWTH PHASES ON TOTAL ALKALOID CONTENT IN *Croton sparsiflorus* MORONG

Aroul Selvam, R.* and K. Manian*

Abstract

***Croton sparsiflorus* Morong, a weed plant, has well adapted in this coastal region and produces alkaloids, during its growth period. The influence of growth phases of total alkaloid content of this species was assessed by estimating the total alkaloid content of various plant parts (root, stem, leaves and reproductive parts) at vegetative and reproductive phases. The results indicate that the alkaloid content was more in all the parts of the plant during reproductive phase as compared to the vegetative phase. Forty percent more alkaloids accumulates during reproductive phase than vegetative phase. The reproductive parts have more alkaloid than other parts of the plants. At both vegetative and reproductive phases similar trend of alkaloid accumulation was noticed in other parts of the plant. The leaves have maximum alkaloid of 8.04 and 6.10 mg/g at vegetative and reproductive phase respectively. Roots of both phases had least amount of alkaloid. The amount of alkaloid in stem at both the phases stands between the root content and leaves content. This is implicit that harvesting of the plant parts during reproductive stage yields more alkaloid of *Croton sparsiflorus* Morong.**

Introduction

India is regarded as paradise of vegetation due to the rich diversity of agroclimatic and socio-cultural conditions prevailing in the country. The World Health Organisation has compiled a list of about 20,000 medicinal plants used in different parts of the globe among 1,99,000 species of angiosperms. This implies that lot of research works are to be aimed to identify the other natural resources for the benefit of mankind. Some of the exotic plants such as *Croton sparsiflorus* Morong., *Parthenium hysterophorus* L., *Prosopis julifora* (S. W) DC introduced through some means were adapted even in the high stressed condition of coastal region of *Pondicherry* and Tamil Nadu. These plants also have sufficient amount of various secondary metabolites having medicinal values. Several studies have shown that alkaloid increases under different kind of stress (Salenz *et al.* 1993). The secondary metabolites in various parts of plants varies considerably due to different growth stages. The most active stage of alkaloid biosynthesis was at the start of flowering in

* Pandit Jawaharlal Nehru College of Agriculture & Research Institute, Karaikal-609 603, Pondicherry.

all the vegetative organs (Blasko and Cordell, 1990). In this background a preliminary study was taken up to find the critical growth phase for extraction of maximum output of phytochemicals in the unexploited plant of *C. sparsiflorus* at the coastal region of Karaikal.

Materials and Methods

The test plants of *Croton sparisflorus* were earmarked in the beginning of early vegetative stage. At active vegetative stage, the plants were harvested and separated into different organs of root, stem and leaves. They were shade dried initially and overdried at 70°C till the plant materials have constant dry weight. After 60 days of first harvest, another harvest was taken up at the time of fully matured reproductive stage of the plant similar to that of first harvest. The dried materials of various organs were meshed in No. 80 sieve and plant extract was obtained with 90 per cent alcohol. The plant extracts of various organs were utilised for estimation of total alkaloid following the method as described by Bose (1984); computed for per plant basis and expressed as mg of alkaloid per plant.

Results and Discussion

The total akaloids of various organs at two different growth phases are tabulated in Table 1. The data revealed that the alkaloid content was more in all the parts of the plants during reproductive phase as compared to the vegetative phase. At vegetative phase, root has significantly lower amount of alkaloid 10.2 and 13.6 per cent lower than the stem and leaves respectively. The fruit has maximum amount of alkaloid among the various plant organs at reproductive phase. This is followed by leaves and stem. The root has least amount of alkaloid during this phase also.

TABLE 1: TOTAL ALKALOID CONTENT IN DIFFERENT PHASES AND DIFFERENT ORGANS OF *Croton sparsiflorus* MORONG.

Plant organs	Total alkaloid content Mg/Plant	
	Vegetative phase	*Reproductive phase*
Root	5.120	7.564
Stem	5.605	7.980
Leaves	6.160	8.040
Fruits	—	8.190
SD	0.520	0.270

The total alkaloid production at reproductive phase was more to the magnitude of 40 per cent than vegetative phase. Higher amount of alkaloid production may be due to alleviate the various stress conditions prevailed in this coastal region. The external factors such as biotic and abiotic stresses such as virus infection and drought influenced the alkaloid content of *Datural innoxa* and *Catharanthus roseus* respectively (Bhakuni and Sudha Jain, 1995 and Selanz *et al.,* 1993). Among the vegetative organs, leaves has maximum alkaloid content both at vegetative and reproductive phases. The leaves of *Hyoscyamus niger* also has higher alkaloid content at the time of flower. (Bhakuni and Sudha Jain, 1995). The belladona leaf possesses maximum amount of alkaloids during early flowering stage, it gets reduced till formation and ripening of fruits (Singh and Chand, 1995). Thus alkaloid biosynthesis varied depend upon the species as well as growth phase of the plant.

The results implicitly indicate that harvesting of the plant part during reproductive stage yields more alkaloid of *Croton sparsiflorus*.

References

Bhakuni, D. S. and Jain, Sudha, 1995. Chemistry of cultivated medicinal plants. *In* : *Medicinal and Aromatic Plants* (eds. K. L. Chadha and Rajendra Gupta.) Malhotra Publishing House. New Delhi. pp. 45-144.

Blasko, G. and G. A. Cordell, 1990. *In* : *The Alkaloids*. Vol.37. (ed. A. Brossi) Academic Press Inc. New York. pp. 1-76.

Bose. R. C., 1984. *Pharmacopoeia Indica*. Bishen Singh Mahendra Pal Singh. Dehra Dun. India.

Singh, Jag Mohan and Ramesh Chand, 1995. Belladonna. *In: Medicinal and Aromatic Plants*. (Eds. K. L. Chadha and Rajendra Gupta) Malhotra Publishing House. New Delhi. pp. 285-295.

Salenz, L., Santamaria, J. M., Vellanueva, M. A., Loyola - Vargas, V. M. and Oropeza, C., 1993. Changes in the alkaloid content of plants of *Catharanthus roseus* L. (Don) as a result of water stress and treatment with abscisic acid. *J. Plant Physiol.* **142** : 244–243.

Recent Trends in Spices and Medicinal Plants Research pp. 125-128, 2000
Ed. Amit Krishna De
Associated Publishing Co., New Delhi

2.10

TRIALLEL ANALYSIS FOR PUNGENCY IN CHILLI

K. P. Sunil* and Rasheed Ahmad**

Abstract

Pungency of chilli is governed by duplicate epistasis (additive x dominance and dominance x dominance) and genetic complementation of genes for pungency having additive effects. The combinations of CA 53 as the first parent, CA 89 as the half parent and CA 582 as the third parent are ideal for the production of triallel hybrids of outstanding performance with respect to higher pungency in chilli.

Introduction

Most of the genetic analysis such as diallel, line x tester and North Carolina plans give information on the relative magnitudes of additive and dominance components of genetic variance. A few of them detect the presence of epistasis. The triallel analysis gives additional information on digenic epistatic interaction components of genetic variances as well as on the effect of the order in which the parents are used in making the triallel hybrids.

Materials and Methods

Sixty triallel crosses were made as per the methods suggested by Rawlings and Cockerham (1962) using 6 varieties of chilli viz. CA 33, CA 53, CA 60, CA 89, CA 484 and CA 582. The expectation of mean square for the cumulative effect of triallel cross model (Ponnuswamy *et al.,* 1974) was followed. Pungency was recorded by estimating capsaicin (per cent) as per Theymoli *et al.* (1982) by reading the absorbance at 650 nm in Spectrophotometer. The genetic components of variances were obtained assuming F = 1 using intermediary components. Statistical analysis were carried out using Indostat—Advance Biometrics package.

Results and Discussion

The analysis of variance indicated significant differences among the triallel cross hybrids of chilli as was seen from highly significant mean square values for pungency

* Department of Agriculture, Mothirakkanni- 680 721, Kerala, India
** Cranfield Biotechnology Centre, Cranfield University, Bedfordshire MK3 OAL, U. K.

(Table 1) The *per-se* performance and magnitudes of components of genetic variances of 60 triallel cross hybrids for pungency in chilli are furnished in Table 2.

TABLE 1: ANALYSIS OF VARIANCE TRIALLEL HYBRIDS FOR PUNGENCY IN CHILLI

Source	*d. f.*	*Pungency (%)*
General line effect of the first kind (h's elim. g's)	5	2.61*
General line effect of the second kind (g's elim. h's)	5	8.80**
Two line specific effects of the first kind (d's elim. s's)	9	2.93**
Two line specific effect of the first kind (d's elim. d's)	19	2.30**
Three line specific effects	21	1.84**
Crosses	59	2.67**
Replication	2	0.13
Error	118	0.35

** P = 0.01 * P = 0.05

TABLE 2: PER-SE PERFORMANCE AND MAGNITUDES OF COMPONENTS OF GENETIC VARIANCES OF TRIALLEL CROSS HYBRIDS FOR PUNGENCY IN CHILLI.

Sl. No.	*Crosses/ Components*	*Pungency (Per cent)*
1	12 x 3	0.71
2	12 x 4	0.64
3	12 x 5	0.47
4	12 x 6	0.53
5	13 x 2	0.63
6	13 x 4	0.73
7	13 x 5	0.68
8	13 x 6	0.73
9	14 x 2	0.66
10	14 x 3	0.76
11	14 x 5	0.72
12	14 x 6	0.72
13	15 x 2	0.58
14	15 x 3	0.70
15	15 x 4	0.90
16	15 x 6	0.70
17	16 x 2	0.87
18	16 x 3	0.77
19	16 x 4	0.80
20	16 x 5	0.75
21	23 x 1	0.60
22	23 x 4	0.63
23	23 x 5	0.61

contd. on page 127

24	23 x 6	0.73
25	24 x 1	0.66
26	24 x 3	0.68
27	24 x 5	0.84
28	24 x 6	0.95
29	25 x 1	0.69
30	25 x 3	0.60
31	25 x 4	0.62
32	25 x 6	0.63
33	26 x 1	0.59
34	26 x 3	0.59
35	26 x 4	0.74
36	26 x 5	0.74
37	34 x 1	0.66
38	34 x 2	0.89
39	34 x 5	0.69
40	34 x 6	0.76
41	35 x 1	0.66
42	35 x 2	0.69
43	35 x 4	0.62
44	35 x 6	0.69
45	36 x 1	0.72
46	36 x 2	0.64
47	36 x 4	0.71
48	36 x 5	0.78
49	45 x 1	0.88
50	45 x 2	0.77
51	45 x 3	0.63
52	45 x 6	0.85
53	46 x 1	0.75
54	46 x 2	0.88
55	46 x 3	0.65
56	46 x 5	0.70
57	56 x 1	0.72
58	56 x 2	0.73
59	56 x 3	0.80
60	56 x 4	0.77
Mean	0.71	
SEm ±	0.03	
CD (P = 0.05)	0.07	
Additive	804.60	
Dominance	–1365.33	
Additive x Additive	–1321.17	
Additive x Dominance	6192.68	
Dominance x Dominance	1543.38	

1-CA 33, 2-CA 53, 3-CA 60, 4-CA 89, 5-CA 484, 6 -CA 582.

Higher values were recorded by triallel cross hybrids for pungency in chilli. Out of the 60 combinations, the fruits of the triallel cross (CA 53 x CA 89) x CA 582 were of highest pungency (0.95%). On the other hand, least pungency (0.47%) was evident in the cross (CA 33 x CA 53) x CA 484. Better performance of triallel cross hybrids over single cross hybrids was reported by Chaudhary (1974) in barley, Sahib and Reddy (1989) in sorghum and Vindhiyavarman (1994) in groundnut. The results revealed that CA 89 was an ideal grand parent for pungency in chilli. The combination CA 53 x CA 89 was found suitable for the improvement of pungency in chilli. In order to establish a desirable third parent to be combined with F_1 hybrids in triallel crosses, CA 582 was superior for pungency in chilli.

In the triallel analysis, non allelic interactions of additive x dominance followed by dominance x dominance types as well as additive gene actions were detected for pungency in chilli. The findings of Bajaj *et al.* (1980), Park and Takatashi (1980) indicated the importance of additive gene effects in governing pungency of chilli. The capsaicin content constitution of triallel cross hybrid chillies are entirely different from that of the single cross hybrid chillies.

As higher proportions of epistatic interactions were detected in triallel analysis, the rapid improvement of pungency through selection may be retarded. Therefore, selection has to be delayed to later generations, when the magnitude on non-fixable interaction effects gets diminished.

References

Bajaj, K. L., Kaur, G. D. and Sooch, B. S., 1980. Varietal variation in some important chemical constituents in chilli. *(Capsicum annuum* L.) fruits. *Veg. Sci.* **7**: 48-54.

Chaudhary, B. D., 1974. Expression of heterosis in single, double and three-way crosses. *Barley Newls.* **17** : 57-58.

Park, J. B. and Takatashi, K., 1980. Expression of heterosis and combining ability for capsaicin content in red pepper. *J. Japan Soc. Hort. Sci.* **49** : 189-196.

Ponnuswamy, K. N., Das M. N. and Handoo, M. I., 1974. Combining ability type of analysis for triallel cross in maize (*Zea mays* L.) *Theor. Appl. Genet.* **45** : 170-175.

Rawlings, J. O. and Cockerham, C. C., 1962. Triallel analysis, *Crop Sci.* **2** : 228-231.

Sahib, H. K. and Reddy, B. B., 1989. Comparison of single, three-way and double crosses in Sorghum. *J. Res. APAO.* **17** : 8-13.

Theymoli, B., Raj, D. Kasthuri, R. and Regaswary, P., 1982. Capsicum and plant characters in chillies. *Indian J. Hort.* **39** : 239-242.

Vindhiyavarman, P., 1994. *Genetic Studies by Diallel and Triallel Cross Analysis in Bunch Groundnut (Arachis hypogae* L. Sp. *fastigiata* Waldron). Ph.D thesis, Tamil Nadu Agric. Univ., Coimbatore, India. pp. 164.

Recent Trends in Spices and Medicinal Plants Research pp. 129-132, 2000
Ed. Amit Krishna De
Associated Publishing Co., New Delhi

2.11

STUDY OF BIO-DIVERSITY AND INTER-RELATIONSHIP IN GENETIC STOCK OF HORSEGRAM (*Macrotylona uniflorum* (LAM) VERD)

S. R. Ramgiry* and P. Raha*

Abstract

An experiment consisted of 25 genotypes of horsegram which were grown during *Kharif* 1995 in randomized complete block design in two replications. The result showed that higher phenotypic and genotypic coefficient of variation were found for branches per plant, seeds per pod and yield per plant. High heritability estimates exhibited for days to maturity, days to 50% flowering and pod length. Significant positive associations were noticed between plant height with branches per plant, pods per plant, seeds per pod and pod length. Similarly pod length with pods per plant, seeds per pod sowing important yield contributing characters in horsegram.

Introduction

India is endowed with a very rich heritage of bio-diversity and two of the world's 18 hot spots of bio-diversity present in India. Rigveda contains detailed information about 2000 medicinal plants (Singh and Pruthi, 1997). Horsegram which is locally known as *Kulthi* is used for curing kidney diseases. However, medicinal plant show low response to genetic manipulation of desirable character (Singh, 1990). Therefore, in the present investigation, diversity and inter-relationship among major yield contributing characters were studied which will be informative for future crop improvement programme.

Materials and Methods

The experimental material comprised of 25 diverse genotypes of horsegram collected from different parts of Jhabua hill zones of Madhya Pradesh, which were grown during *kharif* 1995 at the research field of Zonal Agrilcultural Research Station, Jhabua. Each plot consisted of two rows of 3 meter length keeping 40 cm. row to row and 10 cm plant to plant distances laid out in randomized complete block design with two

*Department of Plant Breeding and Genetics, J. N. Agricultural University, Jabalpur 482 004 (M. P.).

TABLE 1: GENETIC PARAMETERS OF VARIATION FOR DIFFERENT CHARACTERS

		Days to 50% flowering	*Days to maturity*	*Plant height*	*Branches/ Plant*	*Pods/ Plant*	*Seeds/ Pod*	*Pod length*	*Yield/ plant*
1	Mean	63.00	95.50	69.00	9.90	92.10	3.90	4.55	20.40
2.	Phenotypic coefficient of variation	6.15	4.26	8.45	13.52	9.94	13.08	9.04	11.89
3.	Genotypic coefficient of variation	5.96	4.13	7.92	12.80	9.04	11.40	8.63	11.15
4.	Genetic advance as % of mean	7.47	7.89	10.59	2.35	15.58	0.80	0.76	4.45
5.	Heritability	93.90	94.00	88.00	89.60	82.70	75.90	91.20	87.80

TABLE 2: PHENOTYPIC AND GENOTYPIC CORRELATION COEFFICIENTS BETWEEN DIFFERENT PAIRS OF CHARACTERS

Characters		*Days to maturity*	*Plant height*	*Branches/ Plant*	*Pods/ Plant*	*Seeds/ Pod*	*Pod length*	*Yield/ plant*
Days to 50%	P	0.726**	–0.047	0.103	–0.109	–0.182	–0.091	–0.113
flowering	G	0.757**	–0.022	0.113	–0.076	–0.251	–0.098	–0.152
Days to maturity	P		–0.015	–0.053	–0.109	–0.095	0.037	–0.307
	G		0.016	–0.062	–0.120	–0.124	0.040	–0.341
Plant height	P			0.699*	0.807**	0.497*	0.559*	0.270
	G			0.731**	0.918**	0.567*	0.638**	0.313
Branches/Plant	P				0.597*	0.196	0.170	0.282
	G				0.704**	0.185	0.208	0.305
Pods/plant	P					0.482	0.627**	0.294
	G					0.580	0.712**	0.390
Seeds/pod	P						0.610**	0.154
	G						0.697**	0.176
Pod length	P							0.371
	G							0.392

P– Phenotypic correlation coefficient.
G– Genetic correlation coefficient.
* and ** 5% and 1% probability significant level.

replications. All other cultural operations were applied to raise the normal crop. Five plants were selected from each genotypes from every replication for recording detailed observations on days to 50% flowering, days to maturity, plant height, branches per plant, pods per plant, seeds per pod, pod length and yield per plant.

The data were subjected to statistical analysis. The variance and covariance were calculated as per usual procedures. Heritability (in broad sense) and genetic advance as percentage of mean were computed as per formula given by Burton (1952). Phenotypic and genotypic correlation coefficients were estimated using the formula suggested by Miller *et al.,* (1958).

Results and Discussion

The result showed that genotypes differed significantly for all characters under study. The estimates of phenotypic and genotypic coefficient of variability exhibited similar trend (Table 1). The extend of coefficient of variation noticed that branches per plant had the highest genotypic coefficient of variation followed by seeds per pod and yield per plant indicating better scope for genetic improvement. High heritability estimates were found for days to maturity followed by days to 50% flowerings pod length and branches per plant. Similar findings have also been reported by Singh (1990). Further result revealed that branches per plant, seeds per pod and yield per plant had shown higher genotypic coefficient of variation and did not show higher heritability estimates. Similarly, character like days to 50% flowering and maturity possessing low genotypic coefficients recorded higher heritability estimates and resulted in influence of environmental variation in these characters expression.

Estimates of phenotypic and genotypic correlation coefficients were found to have similar trend having higher phenotypic values than the corresponding genotypic values (Table-2), which indicated substantial role of environment in the character expression. Phenotypic as well as genotypic correlation coefficients exhibited significant for 9 out of 28 pairs of character showing poor association. Pod length showed positive strong correlation with plant height, pods per plant and seeds per pod, similarliy between seeds per pod with plant height and pods per plant, pods per plant with plant height and branches per plant. Other positive and significant association between branches per plant with plant height, days to 50% flowering with days to maturity were observed both at phenotypic and genotypic levels. These results are in agreement with finding of Singh (1990). The yield per plant however did not show any significant association with all other characters.

Threfore, it could be concluded from the present study that plant height, pods per plant, seeds per pod and pod length were found to have major yield contributing traits that would give due importance for genetic improvement of horsegram.

References

Burton, G. W., 1952. Quantitative inheritance in grasses. Proceedings 6th International Grassland Congress. **1** : 227-283.

Miller, P. A., Williams, J. C., Robinson, H. F. and Comstock, R. E., 1958. Estimates of genotypic and environmental variances and covariances in upland cotton and their implication in selection, *Agronomy Journal.* **50** : 126-131.

Singh, B. and Pruthi, T. D., 1997. Bio-diversity and Herbal Medicine. Proceedings 6th National Seminar on Biological Diversity and Human Welfare held at Rishikesh 28-30 Sept.

Singh, D. N., 1990. Genetic variability, correlation and path analysis for grain yield and its components in horsegram. *Indian J. Pulses Res.* **3** (1) : 25-30.

Recent Trends in Spices and Medicinal Plants Research pp. 133-139, 2000
Ed. Amit Krishna De
Associated Publishing Co., New Delhi

2.12

TISSUE CULTURE STUDIES IN *Azadirachta indica* AND *Cassia fistula*

S. K. Roy*, P. D. Ghosh* and S. Mandal*

Abstract

Nearly 25-30% of the drugs currently used are still of plant origin. Most of the plant producing these constituents grow wild in the nature, while a few are cultivated. Since such plants have often been collected from the naturally occurring flora, there is an increasing depletion of the natural resources. The present paper highlights the application of tissue culture techniques for rapid multiplication of elite plants through micropropagation of *A. indica.* Meristem cultures leading to multiple shoot formation have been exploited in *Azadirachta indica.* Explants from shoot apices, internodal meristem cultured on MS supplemented with BA and NAA gave better response in shoot multiplication. *Cassia fistula* is one of the important species of *Cassia* which are rich sources of Anthraquinones. Callus tissue obtained from leaf explants were allowed to grow on MS media supplemented with hormones 2,4-D and kinetin. After successful growth of callus. the tissues were harvested after 60 days. Anthraquinones were extracted by ether and ethyl acetate and subjected to TLC with reference to standard commercial anthraquinone.

Introduction

With the progress in the technique of plant tissue and cell culture and its studies of biosynthesis and biotransformation, industrial production of plant medicinal substances by *in vitro* techniques has received more attention in current years (Carew *et al.*, 1965. Puhan, 1971; Ahuja, 1995; Stockigt *et al.*, 1995, Nin, 1997).

The system possesses a number of advantages, for example, it is not subjected to the limitation of soil and seasonal condition and also provides high rate of growth. It has been now possible to industrially produce some plant drugs by cell culture system. It has been demonstrated also that this technique is much superior to the traditional cultivation of plant drug and thus opens up a new way for the production of plant medicinal substances in general.

In the present paper, two techniques such as micropropagation of *A. indica* and callus culture of C. *fistula* with particular reference to anthraquinone production has been discussed.

Azadirachta indica A. Zuss (Meliaceae) is a multipurpose tree for its medicinal and insecticidal properties. It grows in different agroclimatic zones of our country

* Tissue Culture Unit, Cytogenetics Laboratory, Department of Botany, Kalyani University, Nadia, (W. B.).

with a variation in Azadirachtin content (1.5 to 4.5%)from seed carnel, (Ermel *et al.*, 1986). Moreover, traditional methods of genetic improvement through sexual propagation have several limitations (Tomar, 1989). From the literature survey it appears that *in vitro* regeneration of *A. indica* have been made from callus and somatic embryoids and also direct shoot formation from seedling explant (Gautam *et. al.* 1988; Narayan & Jayswal, 1985). To the best of our knowledge there is no report on the *in vitro* cloning methods of adult neem tree. This paper reports methodology of *in vitro* clonal propagation of *A. indica* tree.

Cassia fistula L. (Fabaceae) is one of the important species of *Cassia* which is a rich source of anthraquinones and is widely known for pharmaceutic properties. From perusal of literature (Ahuja and Prasad, 1987) it appears that many useful anthraquinones are present in non-differentiating callus culture of *C. fistula*. The objective of the present programme is to estimate the anthraquinone *in vitro* callus culture qualitatively as well as quantitatively.

Materials and Methods

Neem

Young nodal segments of *A. indica* were collected from the mature trees. Explants of 1.5 cm in length were taken, surface sterilisation was made with 10% teepol for 15 minutes followed by washing with distilled water. Explants were then treated with 0.15% $HgCl_2$ for 5-6, minutes and washed with distilled water for 3 times. Sterilised explants were then inoculated in MS medium supplemented with different combination and concentration of BA, Kin, BAP, NAA and 2,4-D (Table-1). The cultures were then maintained at 26 ±1°C with 16 hrs photoperiod and light intensity of 2000 lux (approx.).

Cassia

Callus Induction

Leaflets were taken as an explant from young branches of *C. fistula* tree. Surface sterilisation was made with 0.2% $HgCl_2$ solution for 5 minutes followed by washing with sterile distilled water. Leaflet explant were then punched from the leaflet lamina with sterile crok borer (5 mm diameter) and single explant was inoculated on solidified MS medium supplemented with 2,4–D, kinetin, BA, NAA, BAP in different combinations and concentrations (Table 2).

Extraction of Anthraquinone

The powdered leaf cells (1gm) from callus was extracted with 30 ml of water at 80°C for 1 hour with frequent agitation. The aqueous extract was decanted and filtered; the mare was re-extracted with another portion of 30 ml water for 30 minutes. The extract was decanted and filtered through same filter paper.

A portion of the aqueous was transferred to a separating funnel, then acidified with 1 drop of HCl and extracted with two successive portion of 40 and 20 ml of ether. The combined ether extract was washed with 2 x 5 ml fraction of water, the

TABLE 1: EFFECT OF HORMONES ON MORPHOGENETIC RESPONSE IN *Azadirachta Indica* A. JUSS.

Name of the media	*Hormone Concentration mg/l*					*Response*			*Remarks*
	2,4-D,	*NAA*	*Kn*	*BAP*	*BA*	*Intensity of Callus initiation*	*Explant showing response (%)*	*Intensity of plantlet*	*(Nature of Callus of plantlets)*
1. MS	2	0	0	1	0	–	–	–	
2. MS	0	2	0	1	0	–	–	–	
3. MS	2	2	0	2	0	–	–	–	
4. MS	0	2	1	0	0	–	–	–	
5. MS	0	1	0	0	1	+	30%	–	callus compact greenish, hard.
6. MS	0	0	1	0	2	–	–	–	
7. MS	0	1.5	0	0	2.5	++	60%	+	callus compact greenish, plantlet regenerates, Plantlets 3–4 branches
8. MS	0	0	1	0	3	+	30%	–	callus white green, compact
9. MS	0	2	0	0	3	++	65%	–	"
10. MS	3	0	1	0	0	–	–	–	
11. MS	0	0	0	2	0	–	40%	++	Plantlets with 2–3 branches
12. MS	0	0	0	3	0	–	20%	+	Plantlets without branch
13. MS	5	0	2	0	0	–	–	–	

+ Poor response, ++ Good response, +++ Excellent response, – No response

TABLE 2: EFFECT OF GROWTH HORMONES ON CALLUS PRODUCTION OF *Cassia fistula* L.

Name of the media	*Hormone Concentration mg/l*					*Response*			*Remarks*
	2,4-D,	NAA	Kn	BAP	BA	Intensity of Callus initiation	Explants showing response(%)	Intensity of plantlet	(Nature of Callus of plantlets)
1. MS	0	0	0	2.5	1	–	–	–	
2. MS	0	0	0	3	1.5	–	–	–	
3. MS	0	0	0	4	2	–	–	–	
4. MS	0	0	1.5	3.5	0	+++	80%	–	Globular, dark brown, compact
5. Ms	0	1	0	4	3	++	60%	–	Globular, greenish friable
6. MS	0	0	1.5	4	0	+	30%	–	
7. MS	1	1	0	0	0	–	–	–	
8. MS	1.5	0	0	3	0	–	–	–	
9. MS	0	1	1	0	0	–	–	–	
10. MS	1	1	0	2	0	–	–	–	

+ Poor response ++ Good response +++ Excellent response – No response ether phase was discarded (free aglycone), in aqueous solution 8 ml of 25% HCl was added and refluxed for 20 minutes. The mixture with new precipitation was extracted with 3 x 30 ml fraction of ethyl acetate. The brown deposits (at the interphase) was taken with the filter, washed with small volume of ethyl acetate.

The series of separators was prepared by 25 ml of ethyl acetate solution to the first and 20 ml of pure ethyl acetate to each of the other two. The three separators were shaken in a consecutive manner with the same two portions of 25 and 20 ml of freshly prepared soaium bicarbonate (saturated) solution. Ethyl acetate portions were discarded and combined bicarbonate solutions were acidified with 10 ml of HCl, the mixture was shaken very gently. Then the mixture was extracted successively with 25 ml and 20 ml of ethyl acetate. The combined ethyl acetate solution was washed with 10 ml of water transferred into 50ml of volumetric flask and completed the volume.

Identification and Determination of Anthraquinone

For TLC, extracts (ether & ethyl acetate) were concentrated and subjected to TLC on Cellulose (Chromedia CC41) alongwith standard anthraquinone and allowed to reach the solvent front 10 cm from base line. Developing mixtures were on butanol : Pyridine : water (6 : 4 : 3) and detected by 15% aniline phthalate in n-but not and leaf at 105°C for 5 minutes. Rf value were determined with formula. (Table-3)

TABLE 3: DETERMINATION AND ESTIMATION OF ANTHRAQUINONE

Solute	*Distance travel by*	*Distance travel by*	*Rf*	*hRf (Rf x 100)*	*Anthraquinone content mg/gm*
Ether extract	2.3	10	0.23	23	0.04
Standard sample	7.9	10	0.79	79	
Ethyle acetate extract	8.4	10	0.84	84	0.06

Results and Discussion

Neem

Plantlets regeneration of Neem explant were initiated within 2-3 weeks of culture on MS basal nutrient media containing BA and NAA. These plantlets grown fully with microshoots of 3-4 in each with 4-6 weeks of culture (Fig. 1). From the experimental results it becomes clear that MS medium supplemented with NAA 1.5 mg/l and BA 2.5 mg/l yielded best result. The results of other treatments have been presented in Table 1. In the initiation of culture there were the formation of compact, greenish callus from the base of the shoot explant. For shoot bud proliferation and shoot growth, MS medium containing NAA (1.5 mg/l) and BA (2.5 mg/l) was found to be the best. NAA in combination with other cytokinin were found inhibitory on proliferation and growth of shoots (Table 1). Hutchinson (1981) and Litz and Jaiswal (1990) were of opinion that BA is an effective cytokin for producing axillary shoot. The superiority of BA over other cytokinin for multiple shoot formation in Mulberry has been reported by Oka and Ohyama (1986).

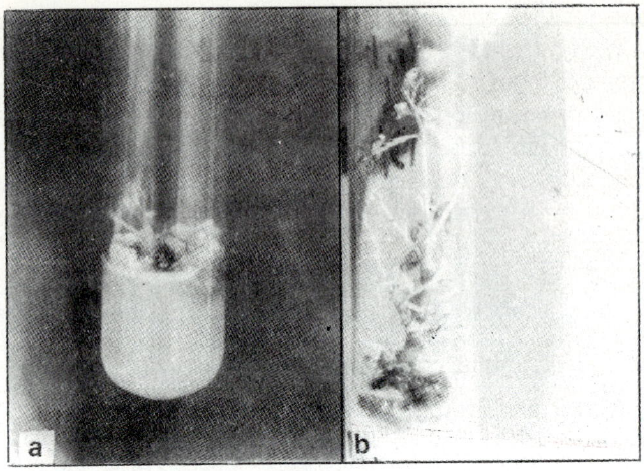

Fig. 1. a. Initiation of plantlet regeneration of *A. indica*
b. Plantlet of same with 2-3 branches

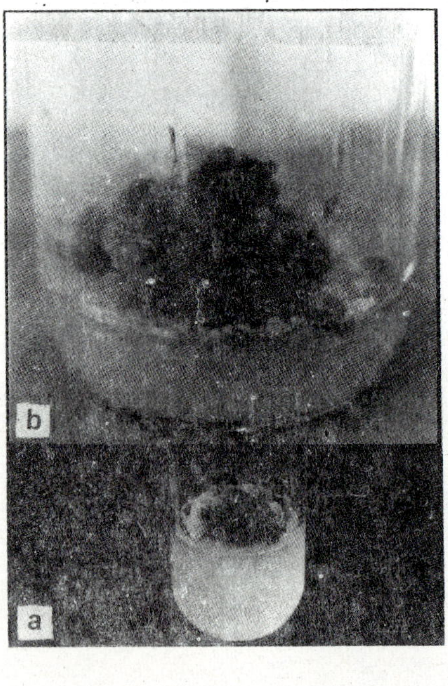

Fig. 2. a. Initiation of callus of *C. fistula* from leaf explant
b. Fully grown friable callus of same

Cassia

Callus induced within 3 to 4 weeks of culture with 2,4-D and kinetin was initially dark brown in colour. The callus was globular and friable in nature Fig. 2. These calli were subsequently subcultured on MS fresh media with 2, 4-D and kinetin in high concentration shows colour variation from greenish to blackish. The growth was very good with the auxin 2,4-D (3.5 mg/l) and kinetin (0.5 to 1.5 mg/l) (Table 2).

It was observed that the extraction of glycoside by following ethyl acetate (0.06 mg/gm) is quantitatively more than ether extraction (0.04 mg/gm) Table 3. This may be due to ether extraction is in free aglycon whereas ethyl acetate is rich in carboxylic anthraquinone.

Acknowledgement

We would like to express our deep gratitude to the Head, Department of Botany, University of Kalyani for providing space and for financial assistance to ARC Govt. of West Bengal.

References

Ahuja *et al.,* 1990. Production of secondary metabolites in plant tissue culture. *Curr. Res. Md. Arom.* Pl., **12** (1) 37-50.

Ahuja et al., 1995. Plants and plant cell cultures as factories for commercial production of high value compounds. *In : Genetic Research and Education: Current Trends and the Next Fifty Years. Indian Society of Genetics and Plant Breeding.* New Delhi. Vol. **IV** P-11655-1665.

Carew, D. P. *et al.,* 1965. Plant tissue culture: Its fundamentals, application and relationship to medicinal plant studies. *Lolydia,* **28** : 1-26.

Ermel, *et al.,* 1986. Azadirachtan content of Neem kernels from different geographical locations and its dependence on temperature, relative humidity and light. *Proc 3rd Int. Neem Conf, Nairobi,* Kenya. pp. 185-194.

Gautam, *et al.,* 1993. Development of shoots and roots in anther derived callus of *Azadirachta indica* A. Juss. A medicinal tree. *Plant cell, tissue and Organ Culture Vol.* **34** : 13-18.

Hutchinson, 1981. Tissue culture : propagation of fruit trees. *In : Proc. COSTED Symp. on Tissue Culture of Economically Important Plants.* Rao, A. N. *(ed.)* Singapore. pp. 113-120.

Litz, *et al.,* 1990. Micropropagation of tropical and subtropical fruits. *In: Micropropagation, Technology and Application.* PC Debergh and Zimmerman RH *(eds.)* Kluwer Academic Publ., Dordrecht, The Netherlands. pp. 247-266.

Narayan, *et al.,* 1985. Plantlet regeneration from leaflet callus of *Azadirachta indica* A. Juss. *J. Tree Sci,* **4** : 65-68.

Nin, *et al.,* 1997. *Agrobacterium* mediated transformation of *Artemisia absinthium* L. (wormwood) and production of secondary metabolites. *Plant Cell Reports* **16** : 725-730.

Oka, *et al.,* 1986. Mulberry (*Morus alba* L.) *In* : Biotechnology in Agriculture and Forestry, Vol. I : Trees I, YPS Bajaj (ed.). Springer Verlag, Berlin, pp. 384-390.

Puhan, S. *et. al.,* 1971. The industrial potential of plant cell culture. *Prog. Indust. Microbiol.,* **3** : 13-39.

Rao, *et al.,* 1988. Induction and growth of callus in *Azadirachta indica* A. Juss. *Crop. Sci.* **15** : 203-205.

Stockigt *et al.,* 1995. Natural products and enzymes from plant cell cultures. *Plant Cell Tissue and Organ Culture.,* **43** : 97-109.

Tomar, 1989. *In vitro* Morphogenesis in some Fabaceous trees *Albizia* spp. *Ph. D. thesis,* University of Delhi.

Chapter 3

Preservation, Processing and Trade

Recent Trends in Spices and Medicinal Plants Research pp. 143-146, 2000
Ed. Amit Krishna De
Associated Publishing Co., New Delhi

3.1

STORABILITY OF CHILLI CULTIVARS UNDER AMBIENT CONDITIONS

A. Nagaraja*, R. C. Jagadeesha, P. K. Basavaraj** & H. S. Yogeesha*****

Abstract

Fruits of twelve chilli varieties harvested during *kharif* 1994, 1995 and 1996 were dried to eleven per cent seed moisture content and stored in paper bags under ambient conditions. Seeds extracted from these fruits were tested for storability after 6,18 and 24 months of storage employing standard germination test (ISTA). Of the twelve, two cultivars KDSC 110-10 and KDSC-6 produced seeds that remained viable for a period of upto 18 months, whereas two others KDSC-510-10 and GPC-80 revealed complete loss of germination. Similar was the vigour of these cultivars thus indicating differential varietal response. Thus, majority of the commercially cultivated chilli cultivars could conveniently be stored under the ambient conditions only till the next sowing season.

Introduction

Chilli (*Capsicum annum* L. Solanaceae) is an important commercial crop of the states of Andhra Pradesh, Karnataka and Maharashtra that contribute to nearly 75 per cent of the country's area and production. The crop is known world wide as an important vegetable cum spice and recently for the extraction of natural dye 'Oleoresine'.

The demand for seed fluctuates in successive years which often results in a large quantity of carry-over stock and the maintenance of high viability in the carry-over seed is of vital importance to the producers (Anuja Gupta *et al.,* 1964). Deterioration during seed storage occurs due to the damage to membrane enzymes, proteins and nucleic acids resulting in the complete disorganization of membranes and cell organelles ultimately causing death of embryo and loss of seed germination (Roberts, 1972). The loss of vigour and viability are common consequences of seed deterioration and the causes and consequences of seed deterioration are still to be ascertained (Roberts, 1983). Doijode (1992) revealed that seed storability and

* Asstt. Plant Pathologist, Main Research Station, Dharwad.
** Asstt. Professors. ARS, Hanumanamatti.
***. Scientist, IARI Regional Station, Dharwad-580 007.
University of Agricultural Sciences, Dharwad-580 005.

seedling vigour were related to storage period in bell pepper. However, Murthy and Murthy (1961) suggested that for seed purpose farmers should store fruits than extracted seeds. The information on storability of different cultivars under ambient conditions is meagre, hence this investigation was undertaken to study the effect of storage of fruits on seed quality over period.

Materials and Methods

The investigation involved 12 popular chilli cultivars KDSC-6, KDSC-110-10, KDSC-210-10, KDSC-510-10, GPC-69, GPC-77, GPC-18,GPC-82, G-3, Jwala, *Byadagi Kaddi* and *Byadagi Dabbi*. All these cultivars were grown at Agricultural Research Station (ARS). Hanumanamatti falling under Zone 8 under the University of Agricultural Sciences, Dharwad during the *kharif* seasons of 1994,1995, and 1996 employing recommended package of practices. In each cultivar, red ripe fruits were picked and dried under sun to 11 per cent seed moisture content (determined by oven method) and were stored separately in brown paper bags under ordinary laboratory conditions at ARS Hanumanamatti till their use. After the last season fruits were stored for 6 months, the seeds from the previous years stored fruits were also simultaneously separated manually, cleaned and subjected to standard germination test as per ISTA (1985). After 14 days of incubation normal seedlings were counted and the root and shoot lengths of 10 randomly selected seedlings were measured (cm). The seedling vigour was computed by multiplication criteria as proposed by Abdul Baki and Anderson (1973). The data was generated for different storage periods and statistical analysis was carried out for 6 months storage data employing CRD.

Results and Discussion

From the Table 1 it is evident that the varieties GPC-87, Jwala, KDSC-6 and KDSC-210-10 recorded significantly higher seed germination. However, all the twelve varieties produced satisfactory levels of germination (60% as per Anonymous, 1988) only from the fruits stored for 6 months. From the fruits stored for 12 months, only two varieties KDSC -110-10 and KDSC-6 revealed acceptable limits while two others GPC-80 and KDSC-510-10 recorded complete loss of germination. From the 24 months stored fruits none of the cultivars showed satisfactory germination. While Thakur *et.al.,* (1989) noticed reduction in seed germination with 5 years storage, Thiagarajan (1994) recorded similar results with 30 months storage. With respect to vigour in terms of root length G-3 was highly vigorous and was on par with KDSC-110-10, but were superior over other cultivars. Vigour in terms of shoot length however, was nonsignificant. With few exceptions both root and shoot lengths reduced with increased periods of storage which is attributed to loss in their germination or delayed emergence.

Vigour index was significantly higher for KDSC-6, and Jwala and decreased with the storage period as also observed by Doijode (1993) in bell pepper. This reduction in vigour index is mainly due to loss in germination. As the vigour index is computed as a product of mean seedling length and germination percentage it is but natural that lesser the germination lesser will be its vigour index.

TABLE 1: QUALITY OF CHILLI SEEDS STORED UNDER AMBIENT CONDITIONS

Varieties/ Months after storage	*Germination*			*Root length (cm)*			*Shoot length (cm)*			*Vigour Index*		
	6	18	24	6	18	24	6	18	24	6	18	24
KDSC-6	81.3	69.0	50.0	3.97	4.33	3.87	9.27	8.63	9.03	1106	864	646
KDSC-110-10	65.3	63.7	44.0	4.66	3.33	3.08	9.27	8.60	7.94	876	760	481
KDSC-210-10	74.7	6.7	4.0	3.40	2.87	2.67	9.55	6.41	5.73	967	62	37
KDSC-510-10	60.0	0.0	0.0	4.16	–	–	9.64	–	–	829	0	0
GPC-69	70.7	32.3	0.0	3.61	2.77	–	8.59	6.90	–	864	313	0
GPC-77	70.7	59.0	15.3	2.92	3.70	3.33	9.72	8.33	7.82	871	710	172
GPC -80	60.7	0.0	0.0	3.15	–	–	8.44	–	–	591	0	0
GPC-82	86.0	37.0	20.0	2.61	2.70	2.86	8.82	8.01	7.15	984	396	179
G-3	60.3	23.3	0.0	5.17	3.32	–	8.81	8.39	–	541	273	0
Jwala	83.0	16.7	0.0	3.74	3.34	–	8.73	7.81	–	1041	186	0
Byadagi Kaddi	70.7	24.3	0.0	3.75	3.82	–	9.67	8.18	–	949	291	0
Byadagi Dabbi	62.0	36.7	0.0	2.92	5.99	–	7.71	9.08	–	493	533	0
F test	S	NA	NA	S	NA	NA	NS	NA	NA	S	NA	NA
LSD(0.05%)	14.3			0.75					230			

NS = non significant; NA = not analysed.

In conclusion it may be said that the chilli cultivars differ in their storability. However, majority of the cultivars retain acceptable limits of germination (60%) and seedling vigour only for a period of 6 months, thus suggesting that for current season's sowing farmers extract seeds from the fruits harvested in the previous season only.

References

Abdul Baki, A. A. and Anderson, J. D., 1973. Vigour determination in soyabean seeds by multiple criteria. *Crop Sci.*, **13** : 630-633.

Anonymous, 1998. *Indian Minimum Seed Certification Standards.* Publ. Central Seed Certification Board, Govt. of India, New Delhi.

Doijode, S. D., 1993. Changes in viability, vigour and certain biochemical aspects during ambient storage of bell pepper seeds. *Haryana J. Hort. Sci.,* **12** (1) : 48-51

Gupta, Anuja, Dharam Singh and Maheshwari, V. K., 1994. Effect of containers on the viability of fungicide treated chilli seeds. *Seed Res.,* **20** (2) : 160-161.

ISTA, 1985. International Rules for Seed Testing. *Seeds Sci. & Technol.,* **13** (2) : 229-355.

Murthy, N. S. R. and Murthy, R. S., 1961. Chilli seed viability in relation to period of storage . *Andhra Agric.J.,* **8** : 246-247.

Roberts, E. H., 1972. *Viability of Seeds.* Chapman and Hall Ltd., London.

Roberts, E. H., 1983. *Causes and Consequences of Seed Deterioration.* Chapman and Hall Ltd., London.

Thakur, P. C., Joshi, S., Verma, T. S. and Kapoor, K. S., 1989. Effect of storage period on germination of sweet pepper seeds. *Capsicum News Letter,* **7** : 58-59.

Thiagarajan, C. P., 1994. Storability of Chilli (*Capsicum annuum* L.). *Madras. Agric. J.*, **81(8)** : 442-445.

Recent Trends in Spices and Medicinal Plants Research pp. 147-150, 2000
Ed. Amit Krishna De
Associated Publishing Co., New Delhi

3.2

FRUIT AND SEED QUALITY OF CHILLI AS INFLUENCED BY DRYING METHODS

A. Nagaraja*, P. K. Basavaraja, R. C. Jagadeesha**, G. B. Shashidhara** and H. S. Yogeesha*****

Abstract

Of the twelve methods employed for drying chilli fruits of the Cv. *Byadagi kaddi*, drying in open air on polythene sheet produced highest germination followed by drying on the ground, cement floor, on plant, cowdung smeared floor and zinc sheet that were at par with each other. On the contrary, all these methods except drying on the plant resulted in significantly higher percentage of whitened fruits. Drying in the oven at 65°C was the best method to minimise percentage of whitened fruits, but resulted in very low germination (12.43%) and thus, not suited to dry fruits intended for seed extraction. Drying on cement floor and on open ground can conveniently be resorted to as they not only produce less number of whitened fruits (28.6 and 26.5%) but also resulted in seeds with good germination (69.98 and 70.42% respectively).

Introduction

Chilli (*Capsicum annum* L.) belonging to family Solanaceae is an important vegetable cum spice crop. Besides, it contains two vital vitamins A and C. Owing to the higher germination and longer storability, Murthy and Murthy (1961) have suggested that, for the seed purpose farmers should store the fruits than the extracted seeds. According to Dhanelappagol (1986), full-red colour fruit produce highest seed germination, field emergence, vigour index, shoot length, root length and seedling dry weight compared to half-red colour fruits and thus suggested that for the seed purpose fruits be picked when fully riped on plant. At this stage, the moisture content in the fruits is so high that they can seldom be stored unless further reduced by drying. However, for large scale seed extraction the harvested ripe fruits are cut or macerated and washed free of pulp and skin and later the seeds are dried in sun on mnesh or screens (Anonymous, 1961). But farmers usually pick ripe fruits and dry them on roof or floor or threshing yard to reduce the moisture content ideal for

* Assistant Plant Pathologist, Main Research Station, Dharwad-5
** Assistant Professors, A. R. S., Hanumanamatti.
*** Scientist, IARI Regional Station, Dharwad-7
University of Agricultural Sciences, Dharwad-580 005, India.

storage or marketing. Unaware of the effects of these local drying methods on the quality of seeds contained inside, the farmers usually select good looking fruits for seed extraction manually. With this background in view an attempt was made to generate information on the effect of different drying methods on fruit quality interalia the quality of seeds contained in them.

Materials and Methods

The drying studies were carried out at the Agricultural Research Station, Hanumanamatti falling in Zone 8 of the University of Agricultural Sciences, Dharwad involving an internationally popular chilli variety *Byadagi Kaddi.* From an healthy crop grown during *kharif* 1996 employing usually agronomic package fully mature red fruits were picked. From this, four kg each of the fruit were subjected to drying by following methods:

M1 On cement floor
M2 On cowdung semeared floor
M3 On zinc sheet
M4 On polythene sheet
M5 On sand
M6 On Mangalore red tiles
M7 Under tree shade
M8 Under house shade
M9 In an oven at 65°C
M10 In open ground
M11 On asbestos sheet, and
M12 On plant (check)

After thorough drying to a seed moisture content of 11 per cent (estimated by oven method) the number of whitened fruits were counted and the percentage worked out. After separating the whitened fruits, good looking fruits were selected and the seeds were extracted manually for use in the laboratory studies at the Seed Science and Technology Laboratory of the College of Agriculture, Dharwad. Standard germination test was conducted as per ISTA (1985). After 14 days incubation, counts were made for the number of normal seedlings. The root and shoot length of ten randomly selected normal seedlings were measured (cm). Vigour index was computed as per Abdul Baki and Anderson (1973). One gram of seeds replicated four times were separately soaked in sterile distilled water over night, the leachate collected and electrical conductivity (EC) was measured using an EC bridge and expressed in μmhos cm^{-1}. The data so generated was analysed employing Completely Radomised Design.

Results and Discussion

Table 1 is indicative of the fact that drying on plant or in oven at 65°C resulted in significantly lower percentage of whitened fruits while drying on red tiles resulted in very high percentage (47.0) of whitened fruits. Among the next best methods, drying in house shade, on open ground and cement floor that were at par appeared useful

TABLE 1: EFFECT OF DRYING METHODS OF FRUIT AND SEED QUALITY OF CHILLI

Drying method	*Percentage of Whitened fruits*	*Germination (%)*	*Root length (cm)*	*Shoot length (cm)*	*Vigour Index*	*EC (µmhos / cm)*
M1	28.6	69.68(56.59)	2.27	7.81	705	74
M2	31.0	66.69(54.73)	2.04	6.31	558	69
M3	33.4	63.04 (52.56)	1.78	7.31	572	59
M4	33.2	72.99 (58.69)	2.43	8.61	805	72
M5	34.5	56.39 (48.67)	1.91	9.01	627	91
M6	47.0	48.66 (44.23)	2.05	8.13	506	112
M7	33.7	56.41(48.65)	2.48	9.05	643	100
M8	25.8	55.12 (47.94)	1.13	6.72	423	71
M9	17.8	12.43 (20.64)	0.38	4.05	60	136
M10	26.5	70.42 (57.05)	1.26	5.34	485	86
M11	39.0	58.03 (49.62)	0.75	6.54	428	66
M12	13.4	68.52 (55.87)	1.61	6.47	560	81
CD (5%)	7.8	7.80	NS	2.83	323	29

NS = non significant

Figures in parentheses indicate transformed values

and practicable methods for large scale drying of fruits that are intended for consumer market, because lesser the whitened fruits higher is the price they fetch. Better fruit quality in drying on plants and in oven were attributed respectively to the natural effects *in-situ* and uniform temperature inside the oven. Contrarily drying on Mangalore red tiles resulted in very high percentage of whitened fruits may be due to higher heat absorbing capacity thus causing scorching injury to the fruit skin.

With respect to the seed quality, oven drying resulted in very low germination (12.43%) and vigour (60), shorter roots (0.38 cm) and shoot (4.05 cm) and high EC (136 μmhos/cm) which is attributed to the stretching of cell membranes due to rapid drying resulting in loss of integrity of seed coat causing death of cells (Ijin, 1957). Contrarily drying in sun on polythene sheet, open ground, cement floor, cowdung smeared floor and zinc sheet produced seeds that were comparable in quality to those obtained from fruits dried *in-situ* on plant. Iljin (1957) also recorded highest germination and vigour with sun drying. Similarly, Dhanelappagol, *et al.,* (1994) also obtained highest germination and vigour in seeds from sun dried fully riped red coloured fruits. However, drying in shade under tree or house caused similar effects as by drying on red tiles or asbestos sheet. This is attributed to the slow drying process that might cause more fungal invasion resulting in large number of decayed or dead seeds in the former in comparison to loss of integrity of cell wall in the latter two methods owing to high temperature effects.

In conclusion, it is said that the fruits intended for seed extraction be invariably dried in sun on any of the locally available materials such as polythene sheet, open ground, cement floor, cowdung smeared floor and zinc sheet (preference decreasing in that order) that produce seeds comparable in quality to those dried *in-situ* on plant, but never on Mangalore red tiles and asbestos sheet or in shade.

References

Abdul Baki, A. A. and Anderson, J. D., 1973. Vigour estimation in soyabean seed by multiple criteria. *Crop Sci.*, **13** : 630–633.

Anonymous, 1961. *Agricultural and Horticultural Seeds–Their Production, Control and Distribution*, FAO, Rome, Italy. pp. 531.

Dhanelappagol, M. S., 1986. Studies on drying of chilli seeds in relation to seed quality. M. Sc. (Agri) Thesis, Univeristy of Agricultural Sciences, Dharwad.

Dhanelappagol, M. S., Shashidhara, S. D. and Kulkarni, G. N., 1994. Effect of stages of harvesting and methods of drying of chilli fruits on seed quality. *Kar. Aagric. Sci.*, **7** (1) : 36–37.

Iljin, W. S., 1957. Drought resistance in plants and physiological processes. *Ann. Rev. Pl. Physiol.*, **8** : 257–274.

ISTA, 1985. International rules for seed testing. *Seed Sci. & Technol.*, **13** (2) : 299–355.

Murthy, N. S. R. and Murthy, B. S., 1961. Chilli seed viability in relation to period of storage. *Andhra Agric. J.*, **8** : 246–247.

Recent Trends in Spices and Medicinal Plants Research pp. 151-154, 2000
Ed. Amit Krishna De
Associated Publishing Co., New Delhi

3.3

CULTURAL PRACTICES FOR CONTROL OF BLIGHT DISEASE OF CUMIN CROP

Kalpna Bhatnagar*, B. S. Sharma* and U. Kant**

Abstract

Cumin (*Cuminum cyminum* L.) belonging to family Umbelliferae is a seed spice of export potential, an important cash crop of India. Blight caused by the fungus *Alternaria burnsi* inflicts losses upto 85 per cent in this crop. In present investigation disease management through cultural practices was attempted by making soil amendment and different date of sowing. Sesame Cake was found effective in reducing the severity (29.33) and increasing yield (311 kg.ha^{-1}) over check (71 kg.ha^{-1}). CODEX was recorded much higher in November sown crop (35.82) as compared to December sown crop (7.14).

Introduction

India being the land of spices, is also the largest producer, consumer and exporter of spices. More than 60 varieties of spices are cultivated in India. The varied agroclimatic zones—tropical, sub-tropical and temperate are quite conducive for cultivation of different types of spices in the country. Rajasthan holds an important position in terms of area and production of spices. The important spices produced in Rajasthan are Coriander, Cumin, Fennel, Fenugreek, Chillies, Garlic, Dill seed and Poppy seed.

Cumin is grown in large area in Rajasthan and it is an important source of income to the farmers. This crop is mostly grown in western districts, such as Jalore, Barmer, Nagaur, Jodhpur, Pali and Sirohi. Besides, it has significant area in Tonk, Ajmer and Bhilwara districts. Cumin is the dried fruit of a small herbacious plant and belongs to family Umbellifereae. This crop is severely damaged by three diseases namely powdery mildew, blight and wilt. These diseases are so devastating that any one of them in its severe form may even completely destroy the crop.

Blight of cumin caused by the fungus—*Alternaria burnsi* (Uppal 1938, Patel 1968, Gemawat and Prasad 1972) inflicts losses upto 85 per cent in this crop. In

*Deptt. of Plant Pathology, A. R. S. Durgapura, Jaipur.
**Department of Botany, University of Rajasthan, Jaipur.

the present investigation disease management through cultural practices was attempted by making soil amendment and different dates of sowing.

Materials and Methods

(a) Effect of soil amendments—A field experiment was laid out in random block design with three replications in each treatment. The experiment was conducted in the 2nd week of November. The organic manure were mixed in soil 25 days before sowing at the rate of 25 quintal per hectare. The soil amendments used were mustard cake, neem cake, cotton, sesame cake, musk melon cake, saw dust, paultry manure along with check. These amendments were finely powdered and thoroughly mixed with soil. The seeds were treated with inoculum before sowing. The disease severity (PDI) and yield data were recorded.

(b) Effect of age of crop and the sowing period on the coefficient of disease index.

The experiment was laid out in random block design with three replications using the suspectible variety CV RS-1 in IM^2 size plot. The crop was shown at 10 day interval from 24th October to 14th December for three years. Inoculum (2.5 x 10^5 per ml. concentration) was applied when the last sown crop was 20 days old.

Disease incidence and severity was recorded after 10th day of inoculation upto flowering in each sown crop. All these observations later on were compiled and CODEX was worked out as suggested by Datar & Mayee, 1981.

$$\text{CODEX} = \frac{\text{PPI} \times \text{PDI}}{100}$$

PPI = Percent disease incidence

PDI = Percent disease severity

Results and Discussion

Disease management was attempted by making soil amendment through oil cakes, saw dust and poultry manure.

TABLE 1: EFFECT OF SOIL AMENDMENT AGAINST BLIGHT DURING THE YEAR 1990-91 AND 1991-92

S. No.	*Treatments 25 Q ha^{-1}*	*Disease severity Av. ϕ value*			*Yield kg ha^{-1}*		
		1990-91	*1991-92*	*Pooled*	*1990-91*	*1991-92*	*Pooled*
1.	Neem Cake	33.77	37.8	35.78	216.67	220.00	218.50
2.	Cotton	40.17	33.25	37.71	263.33	243.33	253.00
3.	Seasame cake	28.65	30.00	29.33	333.33	290.00	311.50
4.	Muskmelon cake	37.84	41.54	39.69	230.00	250.00	240.00
5.	Mustard cake	40.18	40.59	40.39	203.33	230.00	216.50
6.	Paultry Manure	34.73	37.26	36.00	223.33	220.00	221.50
7.	Saw dust	41.94	36.13	39.04	200.00	250.00	225.00
8.	Check	53.96	51.37	52.67	63.33	80.00	71.50
		2.23	2.60	2.79	14.99	25.24	23.57

Pooled data revealed that the effect of different soil amendments was significantly superior over check. However, oil cakes of sesame was found effective in reducing the disease severity (24.33) which was at par with Neem and Poultry manure, over ckeck. Significantly higher yield was obtained by the application of sesame cake (311.50 kg. ha^{-1}) over check (71.50 kg. ha^{-1}). as indicated in Table 1.

However, sesame cake was superior among all, as it brought down the disease severity significantly together with high yield. This might be attributed to the availability of increased dose of nutrition, which made the plant vigor and tolerant against the pathogen. Similarly, Neem cake was found effective in reducing Maiz blight induced by *Drechslera turcia* (Sheelvantar, *et al.* 1986). Effect of different crop growth stages on development of disease was studied. Co-efficient of disease index progressed steadily with the age of the crop irrespective of sowing dates (Table 2).

TABLE 2: EFFECT OF AGE OF CROP AND THE SOWING PERIOD ON CO-EFFICIENT DISEASE INDEX OF CUMIN BLIGHT

S. No.	*Sowing date*	*CODEX (ϕ Value)*			
		1989-90	1990-91	1991-92	Pooled
1.	24th October	28.01	28.18	29.33	28.51
2.	4th November	38.12	34.59	34.75	35.82
3.	14th November	23.86	22.22	22.29	22.79
4.	24th November	18.49	15.18	17.79	17.15
5.	4th December	13.57	13.09	15.82	14.16
6.	14th December	7.58	7.64	6.19	7.14
	SEm ±	0.96	0.53	0.88	0.79
	CD at 5%	3.01	1.66	2.78	2.26
	CD at 1%	4.29	2.37	3.96	3.03

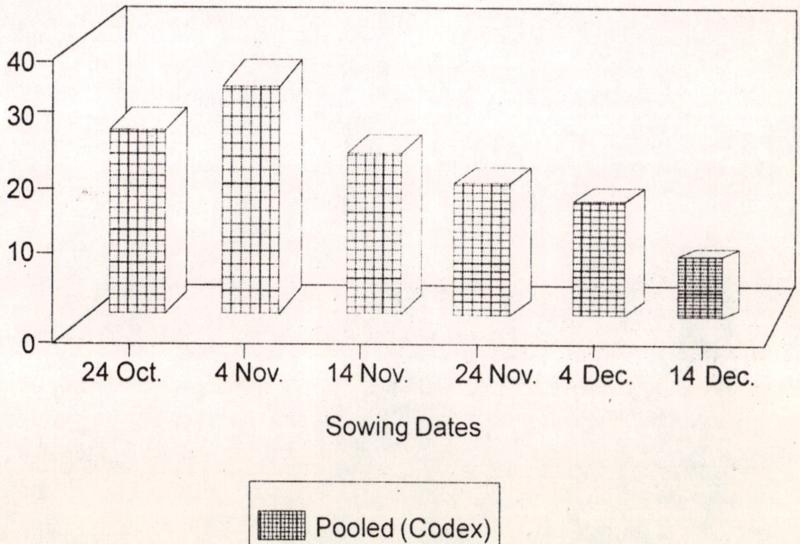

Coefficient of disease index (CODEX) was much higher (35.82) in November sown crop as compared to December sown crop. The least CODEX (7.14) of blight was observed in the crop sown in the second week of December. Inspite of all the suitable meterological parameters for development of disease, higher CODEX was observed in November sown crop that was at flowering stage at the time of inoculation while December crop was at seedling stage that was not the critical stage of the plant for disease development. The present observations are in accordance with that of Patel, 1968, Patil, 1983. However Upal *et al.* (1973), Gaur (1949), Gemawat and Prasad (1972), Solanki *et. al.* (1973) reported that disease symptoms appeared at flowering stage.

References

Datar, V. V. and Mayee, C. D., 1981. Assessment of losses in Tomato yield due to early blight. *Ind. Phytopath.* **34** (2): 191-195.

Gemawat, P. D. and Prasad, N., 1972. Epidemiological stuides of *Alternaria* blight of *Cuminum cyminum. Indian Journal of Mycol. Pl. Pathol.* **2**: 65-67.

Patel R. M., 1968. (Agric.) MSc. Thesis, Sardar Patel University, Anand.

Patel, R. K. and Pillai, S. N., 1983. Age of Crop and sowing period on the incidence of the Cumin blight. *Indian J. of Mycol. Plant Pathol.* **3** 1 : 79.

Uppal, B. M., Patel, M. K. and Kamat, M. N., 1938. Alternaria blight of *Cuminum cyminum* L. *Indian Jour. of Agri. Sci.* **8** : 49.

Recent Trends in Spices and Medicinal Plants Research pp. 155-156, 2000
Ed. Amit Krishna De
Associated Publishing Co., New Delhi

3.4

EFFECT OF AGE ON TUBER YIELD IN SHATAVAR (*Asparagus recemosus* LINN.)

O. P. Singh*, T. P. Singh* and A. L. Yadav*

Abstract

The 65 days old seedlings were planted in size of 6 x 4 m. with spacing 1 x 1 m. Before planting pits were dug in size of 60 x 60 x 60 cm. in the first week of June 1993 and filled with mixture of sand, soil and farm yard manure in ratio of 1 : 1 : 1 about 10 cm. above the ground level in last week of June. Fertilizer were applied at the rate of 40 kg./ha N and 60 kg./ha P205. Observations were recorded after 15, 27, 39 and 51 months of planting on ten randomly selected plants for number of tubers/plant, tuber length, tuber diameter, per tuber weight and fresh weight of tubers/plant. Maximum tuber yield was recorded at the age of 51 months (11.10 kg./plant) followed by 39 months (10.00 kg/plant) and minimum at the age of 15 months 2.50 kg/plant). The optimum tuber yield was harvested from the 39 months old planting.

Introduction

Shatavar (*Asparagus recemosus* Linn.) is a member of family Liliaceae. The importance of *shatavar* was well defined in Ayurveda medicine system. The tubers are bitter in taste and useful in dysentry, inflammation and disease of blood in Ayurveda system (Kirtiker, 1975; Atal and Kapoor, 1982). He further described that in Yunani medicine system tubers are used as laxative, expectorant, galactagogue, tonic and in diseases of kidney and liver.

According to recent chemical investigations, it contains four saponins, shatavar in I-IV and steroid saponins. Saponin mixture from the total alcoholic extract of *shatavar* was found pharmacologically active showing uterine blocking activity.

Shatavar is wildly available in the forest area of U.P., A.P., Orissa, Bihar, M.P., Maharashtra, Kerala and West Bengal. It has wide adaptability. However, the cultivation of shatavar is yet to be brought to the standardization phase. The tubers are the economical part of shatavar. There are too many factors affecting the yield of tubers. Age of the plant is one of the major factor, contributing towards tuber yield. The present investigation was undertaken to workout the effect of age of the plant on tuber yield and yield contributing characters.

*Department of Horticulture, Narender Deva University of Agriculture & Technology, Kumarganj, Faizabad, 224 229.

Materials and Methods

The experiment was conducted at Horticulture Research Station, N.D.U.A.T., Kumarganj, Faizabad. The plants used to conduct the experiment were raised in nursery. The 65 days old seedlings were used in plot size of 6 x 4 m. with spacing 1 x 1 m. pits were dug in a size 60 x 60 x 60 cm. in the first week of June 1993 and filled with mixture of sand, soil and farm yard manure in ratio of 1 : 1 : 1, about 10 cm. above the ground level in last week of June. After filling the pits, plot was irrigated for setting the filling mixture. Before planting seedling N and P205 were applied at the rate of 40 kg. and 60 kg. per ha.

Observations were recorded after 15, 27, 39 and 51 months of planting in the month of October. Ten randomly selected plants from the middle rows were digged to record the data. The data was recorded for number of tubers/plant, tuber length (cm.) tuber diameter (cm), per tuber weight (g) and tuber weight/plant (kg.). For recording the tuber length, tuber diameter and per tuber weight, ten tubers were randomly selected from each plant. The tuber diameter was recorded with the help of vernier calipers.

Results and Discussion

Results of tuber yield and yield contributing character are presented in Table 1.

TABLE 1: PERFORMANCE OF TUBER YIELD AND YIELD CONTRIBUTING CHARACTERS IN *SHATAVAR* (*Asparagus recemosus* Linn).

Sr. No.	*Age of Plant*	*Number of tubers/plant*	*Tuber length (cm)*	*Tuber diameter (cm)*	*Per tuber weight (g)*	*Fresh weight of tuber/ Plant (kg)*
1.	15 months	167	16.27	1.22	14.0	2.50
2.	27 months	264	19.05	1.38	21.0	5.25
3.	39 months	458	19.30	1.44	22.5	10.00
4.	51 months	469	19.60	1.43	23.0	11.10

Maximum number of tubers/plant was recorded after the plant age of 51 months (469) followed by 39 months (458) and 27 months (264) and minimum number of tubers/plant was noted after the plant age 15 months.

Maximum tuber length was recorded at the age of 15 months i.e. 16.27 cm. There were very little differences in length of tubers recorded at the age of 27, 39 and 51 months. Very little difference in tuber diameter and per tuber weights.

Highest fresh weight of tubers/plant was observed after 51 months (11.10 kg) followed by 39 months (10.00 kg.) and 27 months (5.25 kg) and lowest after 15 months (2.50 kg).

From the above results, it can be concluded that optimum tuber yield is obtained after digging of tubers after 39 months.

References

Atal, C. K. and Kapoor, B. M., 1982. *Cultivation and Utilization of Medicinal Plants*. R. R. L., Jammu Tawi, India.

Kirtikar, K. R. and Basu, B. D., 1975. *Indian Medicinal Plants* Vol. **IV**. M/s Periodical Experts, Vivek Vihar, Delhi, India.

Recent Trends in Spices and Medicinal Plants Research pp. 157-161, 2000
Ed. Amit Krishna De
Associated Publishing Co., New Delhi

3.5

PROCESSING AND PACKAGING OF HERBAL/ MEDICINAL PRODUCTS AND ROLE OF APEDA AND ITS ASSISTANCE PROGRAMMES

Vinod K. Kaul*

Abstract

Currently there is a revival of interest in developed nations for natural products. Processing and packaging of herbal products have been discussed in the special emphasis on the role of APEDA.

Introduction

Herbal, medicinal and aromatic plants form an important part of the Indian flora. They find their mention in ancient Ayurvedic texts. The diverse agro-climatic conditions of country have favoured the natural existence and cultivation of a large number of herbs. Generally the scope of the term medicinal and aromatic plants is very wide and the distinction between different terms viz herbs, spices and crude drugs is imprecise. For example, to the American Spice Trade Association the term spice also covers herbs and aromatic seeds. In traditional western usage, the term "Herbs" denotes plants grown mainly in the temperate and mediterranean climates whose leaves, stems and flowering tops are used fresh or dried, mainly for food seasoning. All of us would be familiar with terms like Shilajit, Shikakai, Himax, Teeburb etc.

Although the efficacy of Ayurveda is well known in India, sufficient awareness is still lacking in other countries. However, currently there is a revival of interest in developed nations for natural products. Apart from Ayurveda, Sidha and Unani systems of medicines are existing in our country. All the above systems have their own Medical colleges, hospitals, academic literature base and large number of practitioners in India. Since times immemorial, natural substances derived from herbs have been used with the aim of allaying certain symptoms and maintaining good health till old, as laid down in the Ayurveda—the Science of Life which emphasises the prevention of disease rather than its cure. Ayurveda has stood the

*APEDA, Mayukh Bhawan, Salt Lake City, Calcutta, 700 091.

test of time as provided by the fact that even WHO has recognised it as an alternate system of medicine.

Medicinal plants also have therapeutic use and export potential. Many of the medicinal plants are also recognised by the modern system of medicines and place in various pharmacopes of the world such as USP, BP, EP, DAB, RP etc.

The products based on the Indian medicinal plants consist of powder or whole plants or its parts or aqueous extracts of plants. Many of these plants are food plants and consumed in large quantities as food compared to consumption as medicine for health in few grams or milligrams. These medicinal plants have also established safety and efficacy record which are the most important criteria.

The World Health Organisation has estimated that 80% of the population of developing countries rely on traditional medicines, mostly plant drugs for their primary health care needs. Also, modern pharmacopeia still contain at least 25% drugs derived from plants and many others which are synthetic analogues built on prototype compounds isolated from plants. Demand for medicinal plants is increasing in both developing and developed countries, and surprisingly, the bulk of the material traded is still from wild harvested sources on forest lands and only a very small number of species are cultivated. The expanding trade in medicinal plants has serious implications on the survival of several plant species, with many under serious threat to become extinct.

Medicinal plants form the basis of traditional or indigenous health systems used by the majority of the population of most developing countries. It is a feature of traditional health systems that they span a diverse range of policy areas that extend beyond the immediate domain of health. For example, prospects for the future supply of medicinal plants impact the long term viability of traditional health systems. In many traditional societies, women are the primary herb gatherers and also the herbalists. Of late, there has been a growth of interest in traditional medicine . Unfortunately as habitats for plants disappear and overharvesting for commercial uses reduces stocks of wild medicinal plant material, there is a drop in the availability of plants used as the first and last resort for health care by many rural population in several countries of the world.

Investments are required for the development of appropriate conservation, cultivation and harvesting strategies which will simultaneously meet the demand for low cost and locally available medicines.

Biodiversity and Medicinal Plants

According to FAO there is a loss of tropical rainforests equivalent to 4-5 times the size of the Netherlands each year. These tropical rainforests are the natural resource for medicinal plants used in traditional health systems and of raw material required for modern drug development. It is also estimated that more than 35000 plant species are being used in various human cultures around the globe for medical purposes and many of them are subjected to uncontrolled local and external trade. Natural products from less than 40 tropical species have been incorporated into modern medicine and only a fraction of the tropical flora has been thoroughly analysed for their pharmacological activity. Therefore, the annual extinction rate of around 3000 plant species is a matter of great concern more so since it implies the loss of a potential drug against incurable diseases like cancer and AIDS.

Thus it occurs that even today the plant diversity is still indispensable for human well being and provides all remedies required for health care.

Industrial Utilization of Medicinal Plants

Most developing countries including India are endowed with vast resources of medicinal and aromatic plants which have been used for human welfare in the promotion of health and as drugs and fragrance materials. This close relationship between man and his environment continues even today as a large proportion of people still live in rural areas. A major factor impending the development of the medicinal plants based industries in develping countries has been the lack of information on the social and economic benefits that could be derived from the industrial utilisation of medicinal plants. Except for the use of these plants for local health care needs, not much information has been available on their market potential and trading possibilities. Hence, the economic potential of these plants has not been thoroughly exploited. Also, the demand for medicinal plants in most developing countries has been met by indiscriminate harvesting of spontaneous flora including those in forests resulting in many species becoming extinct. Systematic cultivation of these plants could only be initiated if there is a continuous demand for the raw materials. It is therefore necessary to establish processing facilities preferably in the vicinity of its cultivation so as to create a demand and assure farmers of the sale of raw materials. Thus cultivation and processing should be started simultaneously.

Processing of Medicinal Plants

The promotion and development of processing of plant based products has been given a fresh fillip due to:

a. Green consumerism and the current revival of interest in the use of natural products in developed countries.

b. A growing acceptability of the social responsibility of minimizing socio-economic inequalities in favour of rural people resulting in creating additional job and income opportunities for poor people.

c. Increasing awareness regarding biodiversity conservation and the sustainable and protective use of plant resources.

d. Poor economic conditions in developing countries restricting import, thereby placing increased reliance on medicines using local plant resources.

Some of the constraints associated with the processing of medicinal plants which may result in reducing their competitiveness in global markets are :

- Poor agricultural practices
- Poor harvesting and post harvest treatment practices
- Poor propagation methods
- Inefficient processing techniques leading to low yields and poor quality

- Lack of good manufacturing practices
- Lack of R&D on product and process development
- Difficulties in marketing
- Lack of local market for primary processed products
- Lack of trained personnel and equipment
- Lack of facilities to fabricate equipment locally
- Lack of access to latest technological and market information

Systematic cultivation of many medicinal plants needs specific cultural practices and agronomical requirements which are species specific and dependent on soil, water and climatic conditions. Hence R&D work has to focus on proper selection and identification, propagation methods, cultivation techniques, harvesting, quality control of raw material upto processing stage, post harvest treatment, storage and safety. Organic farming is another practice gaining emphasis worldwide today, farmers have to be trained in all aspects of organic farming of medicinal plants and herbs including obtaining certification from agencies authorised for this.

Process Technology

The medicines for internal use prepared in the traditional manner involve simple methods like hot or cold water extraction, extraction of juice after crushing, powdering of dried material, formulation of powder into paste using water, oil or honey and even fermentation after adding sugar.

Traditional herbal medicines have been produced by use of primitive methods which have many disadvantages and could be corrected by selecting the suitable technologies. This would ensure quality and would also be profit oriented. The question arises whether the introduction of modern technology for the production of traditional medicines will be acceptable to the practitioner who has sole faith in extemporaneous preparations. This situation can be tackled by proper guidance, counselling and education whereby the disadvantages of the old methods and the advantages of the new methods can be imparted.

Processed products (Galenicals) from plants could be standardised fluid/solid extracts or powders or tinctures. Standardised extracts of many plants e.g. *Aloe* species, *Atropa belladona*, *Cassia angustifolia*, *Capsicum annum, Centella asiatica, Cephaelis ipecaduanha, Digitalis* species: *Commiphora mukul*, *Panax ginseng* etc are widely used in health care. Some of these have to be formulated for incorporation into modern dosage forms. New formulations require some development work, particularly on account of the nature of the processed product. Plant extracts are difficult to granulate, sensitive to moisture and prone to microbial contamination. Hence the types of expedients to be used and the processing parameters have to be determined.

New Technologies

Improved methods for the processing of medicinal and aromatic plants and new techniques for quality assessment are being developed rapidly and continuously and they have to be introduced to developing countries if they are to forge ahead to keep up with recent developments and new global requirements. One such method of recent development is Supercritical Fluid Extraction of natural products as an alternative for solvent extraction. It is now used both for processing of phytopharmaceuticals and other plant products and for the removal of trace amounts of contaminant substances such as pesticides, toxins and surfactants.

Quality Assurance

The control of the quality of the raw material, finished products and of processes is an absolute necessity to produce goods for global markets and human consumption. International Standard Specifications exist for some processed products and some countries and buyers have their own requirements. The quality requirements for medicinal plant preparations are stringent in terms of content of active principles and toxic materials.

Quality has to be built into the whole process beginning from the selection of propagation material to the final product reaching the consumer. It is, therefore, a management system where all steps involved in the industrial utilization process have to be properly and strictly controlled to produce the desired quality products. All elements of Total Quality Management (TQM) have to be introduced in any industrial project. The requirement of ISO-9000 certification and Good Manufacturing Practices (GMP) have to be introduced and the personnel trained so that enterprises could introduce the proper systems needed for certification.

In summary, it implies that the development of any industrial processing activity in medicinal plants has to be linked to the specific needs, socio-cultural background, resource potential and the technological capabilities of each country.

Packaging of Medicinal Plants

As far as packaging is concerned it depends on the type of processed medicinal plants i.e. as crude drugs and essential oils they can be packed in plant containers, glass bottles etc. Since by the very mandate APEDA monitors international trade for herbal and medicinal plants and the crude drugs and essential oils are being handled by CHEMEXIL, packaging details are generally available with CHEMEXIL.

Recent Trends in Spices and Medicinal Plants Research pp. 162-173, 2000
Ed. Amit Krishna De
Associated Publishing Co., New Delhi

3.6

PRODUCTION AND EXPORT OF SPICES FROM INDIA

C. K George*

Abstract

An overview of the production and exports of different spices of India has been discussed in detail.

Introduction

India is the foremost country in production, consumption and export of spices. Blessed with tropical, sub-tropical and temperate climates and varied soil types make India possible to grow almost all spices. There are 26 States and 5 Union Territories in India. Every State and Union Territory grows at least a few spices. Andhra Pradesh, Karnataka, Kerala, Gujarat, Madhya Pradesh, Maharashtra, Orissa, Rajasthan, Tamil Nadu and West Bengal are the leading spice producing States. North Eastern Region and Andaman and Nicobar Islands are potential areas for spice cultivation. The important spices produced in India are as under:

Major spices: Black pepper, cardamom, ginger, turmeric, chilli and garlic.
Seed spices: Coriander, cumin, fennel, fenugreek, celery, bishop's weed, dill and caraway.
Tree spices: Clove, nutmeg, cinnamon, allspice, tamarind, kokum, garcinia and curry leaf.
Herbal Spices: Rosemary, thyme, sage, savory, marjoram, oregano, basil and parsley.
Other spices: Saffron, vanilla, mint and horse-radish.

Total production of spices in India during 1996-97 was around 29.37 lakh tonnes from an area of 25.40 lakh hectares. The production of individual spices during 1996-97 is given in Table 1.

Bulk of the spices production is consumed within the country. They are used in various ways of which culinary preparations are the most significant. Indian curry is rich in flavour and aroma. In each curry preparation, a few spices are used to make

*Executive Director, Spices Board, Cochin- 682 025.

it a gourmet's dish. Spices find applications in many Ayurvedic medicines which are popular in India. Extracts of spices are used for perfumes, cosmetics and beauty care products. In the religious rites some of the spices have a prime place. Almost 93% of the production is thus consumed in the country leaving the rest for export.

TABLE 1: ESTIMATED PRODUCTION OF SPICES DURING 1996-97

Spice	*Production (Tonnes)*
Pepper	65, 000
Cardamom (Small)	6, 625
Cardamom (Large)	5, 150
Chilli	800, 000
Ginger	250, 000
Turmeric	532, 000
Garlic	460, 000
Coriander	225, 000
Cumin	180, 000
Fennel	27, 000
Fenugreek	47, 000
Clove	2, 200
Nutmeg	4, 000
Cinnamon	500
Tamarind	300, 000
Kokum	10, 000
Camboge	10, 000
Herbs	10, 000
Others (Dill, Aniseed, Saffron, Vanilla etc)	3,000
Total	29,37,475

Development Policy

Agriculture being a State subject, production of spices barring cardamom is the responsibility of State Governments. Cardamom, a plantation crop exempted from land ceillng laws, is under the purview of Spices Board, Ministry of Commerce, Govt. of India. The State Governments provide funds for developing spices other than cardamom from annual budgets. However, the major development programmes are financed by centrally sponsored/central sector schemes of the Ministry of Agriculture, Govt. of India. Indian Council of Agricultural Research (ICAR) under the Ministry of Agriculture, has set up Indian Institute of Spices Research for research on spices in the important producing State, Kerala. Most of the Argicultural Universities (Total 29 including one central university) also carry out research depending upon the importance of spices grown in the States. The ICAR is having an All India Co-ordinated Spices Improvement Research Project with 20 centres involving 15 Agricultural Universities. Twelve important spices, namely, black pepper, small cardamom, ginger, turmeric, coriander, cumin, fennel, fenugreek, celery, nutmeg, cinnamon and large cardamom have been brought within this project for co-ordinated research. The Spices Board has a research establishment mainly for research on small and large cardamoms and also on selected spices including collection of germplasm, pureline selection, hybridisation, introduction of new spices, assessment of fertilizer requirements, development of control measures for pests and diseases, formulation of integrated pest and disease management, water management, use of biotechnology, streamlining organic farming and improvement of post harvest management.

Strategy for Development

Indian spices have generally low productivity but possess high intrinsic quality. Extensive surveys and studies have been carried out to identify the reasons for low productivity. Lack of enough programmes for large scale production and meagre distribution of high yielding planting materials of recommended varieties remain to be the foremost factors for the low yield. Non-manuring, under manuring or imbalanced manuring and non-adpotion of improved cultural practices are certain other constraints.

Pest and disease incidences and reduced use of chemical control measures also affect production. In the case of pepper and ginger, Foot-rot and Soft-rot diseases take a heavy toll of crops every year. Fruit rot of chilli and Wilt diseases of cumin, coriander and fennel also adversely affect the productivity considerably. Plant protection and control measures for pests and diseases are costly and hence not adopted by many farmers.

High production technology developed at the Research Institutes have largely not been transferred to the field due to insufficient extension personnel and lack of proper infrastructure. Efforts to motivate farmers to adopt the newly developed technologies through field demonstrations have not been carried out adequately. High cost of inputs and lack of enough government incentives also hinder crop development. Spices growers being mostly small and marginal farmers are not financially sound to adopt new technology. Training of farmers on production, protection and processing technology are required to be carried out in a big way. Marketing of spices ensuring remunerative process also needs improvement.

Though spices are important cash crops and required to meet daily needs in the preparation of food of the poor and rich as well, not much attention was paid on these crops till the end of the VII Five Year Plan. The VII Five Year Plan which started in 1991-92 had set apart a larger outlay of Rs. 125 crore for various programmes for spices development by the Ministry of Agriculture. As per a provisional estimate, production of all spices together during 1990-91 was 20.3 lakh tonnes. An overall growth rate of 8% was envisaged during this Plan till 1996-97. The major spices covered under this programme were black pepper, ginger, turmeric, chilli, coriander, cumin, garlic, fennel. fenugreek celery, tamarind, clove, nutmeg, cinnamon and saffron. For developing small and large cardamom an amount of Rs. 21.7 crore was provided by the Spices Board. The Central and Centrally sponsored schemes under the Ministry of Agriculture was for the entire country covering 25 States and 2 Union Territories.

Existing and Upcoming Programmes

Supply of quality planting materials, distribution of mini kits, laying out of demonstration plots, replanting, introduction of irrigation facilities,improvement of post harvest management and modernising warehousing facilities are the programmes generally taken up for different spices.

Pepper being an important spice more than 50% of the allocation has been for this crop. The component programmes taken up for pepper include production of planting materials, rehabilitation of old pepper gardens, distribution of input kits, protection against Foot-rot disease, eradication of Little Leaf disease, laying out of fertilizer demonstration plots, demonstration of high production technology and area expansion. For ginger, turmeric and chilli, production of nucleus seed material, laying out of demonstration-cum-seed multiplication plots, distribution of mini kits and area expansion were the important programmes taken up. For improving the appearance of turmeric, polishing drums were distributed. *Paprika* development was initiated by supplying of seeds as well as demonstrating the cultivation technology. Distribution of planting materials, establishment of scion nursery centres and conduct of demonstrations on better cultivation practices were the programmes for the development of tree spices such as clove, nutmeg, cinnamon and tamarind.

Attention was paid for increasing the production of seed spices such as coriander, cumin, fennel, fenugreek and celery. Production of nucleus seeds, laying out of demonstration plots and distribution of mini-kits were the measures identified for these spices. In garlic, production of breeder seed, foundation seed and certified seed, besides field demonstrations were the programmes implemented.

Saffron is grown in India in Jammu & Kashmir. Establishment of corm (Planting material) multiplication plots and demonstration of better cultivation practices are the programmes implemented for this crop. Efforts are also made to introduce its cultivation in non-traditional areas. For large and small cardamom development, planting materials are distributed and farmers are encouraged to take up replanting of old and diseased plantations and introduction of irrigation. According to the specific needs of certain spices a few general programmes are also implemented in the growing areas. Such programmes are supply of plant protection equipments, promotion of soil conservation measures, introduction of modern marketing and setting up of farmers co-operatives. By the implementation of these programmes it is envisaged that the production of various spices in India will increase to 29.35 lakh tonnes by the end of the VIII Plan period.

Ministry of Agriculture and the Spices Board have reviewed programmes implemented in the VII Five Year Plan and decided their modification or continuance based on the field experience. A reformulated strategy for development of spices has been evolved by the Planning Commission at the end of this exercise. The IX Five Year Plan which commenced in 1997-98 have directions to substantially increase production by improving productivity. Efforts are also being made to cultivate varieties with better intrinsic quality. The hygienic quality of all the spices will be improved and farmers encouraged to use only safe pesticides which have permissible limits or leaving no residue of pesticides. The IX Five Year Plan which will end in 2001-2002 is proposed to have annual average growth rate of 8% in production. The targets arrived at for production of spices by the Working Groups for IX Five Year Plan are given in Table 2.

The production of all spices together will be increased to 43.11 lakh tonnes by the end of IX Five Year Plan i.e. 2001-2002 from the expected level of 29.37 lakh tonnes during 1996-97 which is the last year of the VIII Five Year Plan.

TABLE 2: ESTIMATED PRODUCTION OF SPICES DURING 2001-2002

Spice	*Production (Tonnes)*
Pepper	120,000
Cardamom (Small)	10,000
Cardamom (Large)	8,000
Chilli	1200,000
Ginger	400,000
Turmeric	700,000
Garlic	700,000
Coriander	330,000
Cumin	270,000
Fennel	40,000
Fenugreek	80,000
Clove	5,000
Nutmeg	7,000
Cinnamon	1,000
Tamarind	400,000
Kokum	12,000
Camboge	12,000
Herbs	12,000
Others (Dill, Aniseed, Saffron, Vanilla etc)	4,000
Total	43,11,000

Adoption of Modern Technology

Because of the intensive research carried out both on production and processing a number of findings have been made including measures to reduce the cost of production and to improve both intrinsic and extrinsic quality. Modern approaches such as biotechnology, harvesting solar energy, economic use of water through drip irrigation and improved post-harvest technology both at farm and factory levels have been developed. Biotechnological methods such as micropropagation techniques have been standardised for spices like cardamom, vanilla, pepper, ginger, turmeric, saffron, tamarind, curry leaf and a number of herbal spices. Special programmes are under implementation in some of the tissue culture laboratories for evolving protocol for micro-propagation of clove, nutmeg, kokum and camboge. *In-vitro* tissue culture of costly spices such as saffron and mace has been attended to in a few laboratories. Planting materials of small cardamom and vanilla produced through micropropagation technique have already been distributed to the growers on large scale. Drippers and sprinklers which are free from operational defects are made in India and being used for irrigating selected spices.

Use of plastics is encouraged in production and processing. Plastic sheets for mulching fields for weed control and water conservation are being popularised under a special programme. While plastic green houses are erected for easy nursery management, multi-layered black polythene sheets are distributed for drying different spices. Programmes are underway to construct modern warehouses in the production centres all over the country.

Processing research has been taken up by the Central Food Technological Research Institute and Regional Research Laboratories under the Ministry of Science & Technology. Many technologies have been developed as a result of it for making value added products from spices. Private laboratories are also playing a significant role in developing new products considering the changing food habits within the country and abroad. Spices yield oils and/or oleoresin and by setting up a series of extraction units, India now stands as the number one country in the preparation of these products for export.

Spices can be processed into a number of value added products other than oils and oleoresins. Black pepper, on processing, yields products which can improve appearance and quality of food. Such products are white pepper, dehydrated green pepper, pepper in brine, frozen pepper, freeze dried pepper, pink pepper and sterilised pepper. Bleached and unbleached ginger, ginger candy, ginger cube, ginger preserve, ginger titbits and ginger syrup can be prepared from fresh ginger. The oil extracted from cardamom is used for making cardamom cola, cardamom biscuits, cardamom toffee, and cardamom flan. A few companies in India are already making these products and selling them here and abroad.

Quality Considerations

Quality is presently the watch-word in the international trade. USA, Europe and Japan have fixed different hygienic standards for spices. Accordingly spices should meet specifications prescribed by these countries in respect of physical and microbial contaminations, filth, pesticide residue and aflatoxin. Even though the quantity of spices in the daily intake of foods is negligible, some of the specifications prescribed are very stringent. The Spices Board is taking comprehensive steps to ensure that only quality spices are exported from India.

The Board has set up Quality Evaluation and Upgradation of Laboratory to monitor the quality of the spices produced in different parts of the country. This laboratory is well equipped to analyse spices and spice products from contaminants like physical, chemical, microbial, pesticide residues, aflatoxin, solvent residues etc. The laboratory identifies quality problems in the spice industry by undertaking surveys during production, harvest and post-harvest operations.

The Board helps the exporters for improving the facilities available with them for quality check of spices. Exporters are encouraged to buy modern analytical equipments, so that they can test the quality of spices before export. Spices Board logo is provided for consumer packs of less than 1 kg and also for institutional packs upto 25 kgs. Spice House Certificates are given for those having clearing and grading equipments and modern warehouses. So far 14 consumer packaging units have been recognised under logo scheme and 44 exporting units given Spice House Certificate.

Training of farmers, traders and exporters has been identified as the essential input for improving the quality of spices handled at various stages. A core group of faculty members has been formed for this purpose by taking experts from Spices Board, Export Inspection Agency, Directorate of Marketing and Inspection and State Agricultural Universities. Master trainers are made by imparting training to the officers of the State Agriculture/Horticulture Departments in the important spice growing States. They train the junior level agricultural officers and farmers on production of quality spices taking into consideration of the requirements of individual spices. So far 4939 officers of the Agriculture/Horticulture Departments have been trained as master trainers and 1,48,574 farmers on ensuring quality during the last 10 years. Various extension materials on the quality maintenance of black pepper, ginger, turmeric, fennel, cumin and chilli have been distributed to the master trainers as well as farmers.

Exporters being the key personnel in the international trade a series of training programmes and workshops have been educated on the quality requirements of spices in the importing countries and the steps to be taken to attain that quality. Table 3 gives year-wise trainings conducted for officers, farmers, traders/exporters since the inception of the Spices Board.

TABLE 3: QUALITY IMPROVEMENT TRAINING PROGRAMMES

Year	*Officers of the State Agri./Horti. Deptt.*	*Farmers*	*Traders/ Exporters*
1988-89	—	1100	—
1989-90	100	18000	—
1990-91	300	10000	—
1991-92	360	22020	—
1992-93	394	10400	—
1993-94	252	11460	—
1994-95	576	16910	45
1995-96	762	23945	43
1996-97	295	31739	83
1997-98 (Estimated)	1900	43000	575
Total	4939	148574	746

The services of NGOs are also utilised for the training of farmers. At present, 5 NGOs are closely involved in the training programmes of the Board. Emphasis is given for training women farmers and workers as a number of agricultural operations in spice production and processing are attended to by them.

India—Spice Basket

As per the terminology of the International Organisation for Standardization there are 109 spices. These spices are grown in various countries. India with diverse and favourable agro-climatic conditions is able to grow almost all spices. The well established research and development set up in the country supports not only for increasing production and quality of traditional spices, but also for introduction of new spices from other countries and area expansion. The strong production base of spices developed through substantial domestic consumption has made the Indian spice industry on firm footing. The country is now able to supply increased quantity of a variety of spices meeting specific quality specifications of the importing countries.

Export of Spices

The quality of Indian spices, their captive flavour and aroma and exquisite nature have impressed discerning consumers all over the world. More than 120 countries import Indian spices and consumers buy them as they want the best. The growing preference for Indian spices in the world market is reflected by the increasing demand for Indian spices compared to other countries combined. During 1980-85,

India was having a share of 27.5% in the world trade in spices, which has increased to over 45% in 1996-97.

Export of spices from India has shown a very impressive increasing trend during the last 4 decades. In 1960s, India's average export was around 52,500 tonnes valued at Rs. 21.7 crore. In 1970s this was increased to 70,840 tonnes valued at Rs. 83.1 crore. In 1980s the average export again went up to 85,270 tonnes for a value of Rs. 203.6 crore. As per the latest figures during 1996-97 India exported 2,19,400 tonnes of spices for a value of Rs. 1180 crore. Itemwise export of spices from India during 1996-97 is given in Table 4.

TABLE 4: ITEM-WISE EXPORT OF SPICES FROM INDIA DURING 1996-97
(Qty. in Tonnes, Value in Rs. Crore)

Item	*Quantity*	*Value*
Pepper	47,770	416.52
Cardamom (Small)	240	9.21
Cardamom (Large)	1,450	10.47
Chilli	51,900	209.64
Ginger	28,350	57.24
Turmeric	21,600	54.71
Coriander	12,700	30.81
Cumin	4,500	25.08
Celery	3,400	7.02
Fennel	3,800	13.70
Fenugreek	8,100	10.74
Other seeds (1)	2,500	6.05
Garlic	3,780	6.28
Other seeds (2)	20,920	43.73
Curry Powder	4,250	18.23
Mint Oil	2,120	104.90
Spice Oils & Oleoresins	2,020	155.67
Total	219,400	1180.00

(1) Include Ajwanseed, Dill Seed, Poppy seed, Aniseed etc.

(2) Include Tamarind, Asafoetida, Cinnamon, Cassia, Tejpat, Kokum, Saffron etc.

Pepper

Indian pepper is well-known all over the world. It is considered to be superior in aroma and pungency compared to that of other countries. Indian pepper grade which is mainly traded in Malabar Garbled Gr.I. Tellichery Bold and Tellichery Extra Bold pepper are other important grades. India exports both black pepper and value added products in sizable quantities. The export of pepper which was 37,083 tonnes valued at Rs. 200.3 crore in 1986-87 has increased to 47,760 tonnes valued at Rs. 416.5 crore in 1996-97. The important buyers of Indian pepper are USA, Canada, France, Germany, Italy, Netherlands, Poland and Singapore. Indonesia, Vietnam, Malaysia, Brazil, Thailand, Sri Lanka and China are other important producers competing with Indian pepper.

Cardamom (Small)

India was the largest producer and supplier of small cardamom until 1979-80. Since then Guatemala overtook our country both in production and export. But, Guatemala accounts for 90% of the global trade which is estimated to be around 14,000 tonnes. In 1985-86, India exported 3272 tonnes of small cardamom valued at Rs. 53.5 crore. Exports came down to just 190 tonnes valued at Rs. 7.5 crore by 1992-93. Though export increased to 500 tonnes valued at Rs. 12.4 crore in 1995-96 it came down again to 240 tonnes valued at Rs. 9.2 crore in 1996-97. Increased use of cardamom within the country and the prevailing high domestic price are the main reasons for poor export. Saudi Arabia, Kuwait and Japan are the major buyers of small cardamom.

Cardamom(Large)

Large cardamom is the most important spice grown in Sikkim and Darjeeling district of West Bengal. Nepal and Bhutan are the other producers of large cardamom. A quantity of 195 tonnes valued at Rs. 0.96 crore was exported in 1986-87. Export has increased substantially from 1990-91 onwards to over 900 tonnes and reached 1450 tonnes valued at Rs. 10.5 crore in 1996-97. Pakistan, Afghanisthan, Singapore, UAE and UK are the important buyers of large cardamom.

Chilli

Though commercial chilli is an introduction by the Portuguese, India has become the world's largest producer since last many years. Pakistan, China and Bangladesh are the other producers of chilli. As per the official statistics for 1992-93 production of chilli in India was a record at 9,30,000 tonnes. Chilli export from India has shown remarkable growth during the last few years. Export which was 4327 tonnes valued at Rs. 8.3 crore in 1986-87 increased to 56,073 tonnes valued at 194.1 crore in 1995-96 creating a record. Though quantity of export has come down in 1996-97 to 51,900 tonnes value realised has increased to 209.6 crore. Bangladesh, Sri Lanka, Singapore, Malaysia and USA are the important buyers of Indian chilli.

Ginger

India is the largest producer of ginger contributing over 30% of the world production. China, Bangladesh, Nigeria, Sierra Leone and Sri Lanka are other important producers of ginger. Exportable varieties of ginger are mainly cultivated in Kerala. Export of ginger from India has shown an increasing trend. During 1986-87 export was 4,843 tonnes valued at Rs. 5.7 crore. It has increased to a record level of 28,350 tonnes valued at Rs. 57.2 crore in 1996-97. USA, Pakistan, Bangladesh, Saudi Arabia and Yemen Arab Republic are the major importing countries of Indian ginger. Japan imports annually around 35,000 tonnes of ginger in brine from neighbouring countries.

Turmeric

India is the largest producer and supplier of turmeric in the world. China, Peru and Thailand are the other producers. India shares about 90% of the world trade of turmeric. Varieties cultivated in southern Kerala and in the north-east India are reported to be superior in quality with high curcumin content. Export of turmeric from India has shown an increasing trend from 1987-88 to 1994-95. Accordingly, the exports increased from 8,747 tonnes valued at Rs. 9.2 crore in 1987-88 to the record level in quantity of 28,286 tonnes valued at Rs. 45.2 crore in 1994-95. The quantity exported marginally declined in 1995-96 and further to 21, 600 tonnes in 1996-97. However, value of export in 1996-97 increased to Rs. 54.7 crore. Iran, UAE, Japan, UK, USA and Singapore are the major markets for Indian turmeric.

Garlic

World production of garlic is estimated around 3 million tonnes. India contributes about 11% of the world production. Korea, Spain, Thailand and Egypt are the other major producers of garlic. Export of garlic was only 574 tonnes valued at 0.5 crore in 1986-87. 1991-92 witnessed the highest export of 10,282 tonnes valued at Rs. 8.3 crore. During 1996-97, export was only 3780 tonnes valued at Rs. 6.3 crore. Bangladesh, Japan and Mauritius are the important buyers of garlic from India.

Coriander

India has premier position for the supply of coriander to the world market. Coriander export was only 1,177 tonnes valued at Rs. 1.6 crore in 1986-87. Export has increased especially in the last 5 years. During 1996-97, a quantity of 12,700 tonnes was exported with a value of Rs. 30.8 crore. Malaysia, Singapore, South Africa and UAE are the leading buyers of Indian coriander.

Cumin

India is the major producer of cumin. Export during 1986-87 was only 1,761 tonnes with a value of Rs. 3.4 crore. It has increased to more than three times in 1994-95 and reached 5,618 tonnes valued at Rs. 24.0 crore. However, in the subsequent years export has been dwindled and reached 4,500 tonnes valued at Rs. 25.0 crore in 1996-97. Nepal, Pakistan, USA, Sri Lanka, Algeria, Canada, and Japan are the major buyers of cumin.

Fennel

Export of fennel has increased in the last 10 years. As against 864 tonnes valued at Rs. 1.2 crore in 1986-87, export was 3,800 tonnes valued at Rs. 13.7 crore in 1996-97. Canada, South Africa, Saudi Arabia, Singapore, USA, UK and UAE are the important buyers of fennel.

Fenugreek

Export of fenugreek has increased in the last 5 years. The record export of fenugreek was in 1995-96 at 15,135 tonnes valued at Rs. 18.7 crore. It has come down to 8,100 valued at 10.7 crore in 1996-97. Bangladesh, Israel, Morocco, Malaysia, Netherlands, South Africa, Sri Lanka, Saudi Arabia and UAE are the main buyers of fenugreek.

Celery

Though celery is almost a monopoly produce of India, export has not been growing steadily..The record export in quantity was in 1993-94 at 4,130 tonnes valued at Rs. 6.6 crore. In 1996-97, quantity of export came down to 3,400 tonnes, but value increased to Rs. 7.0 crore. Canada, France, Germany, USA and UK are the regular buyers of celery from India.

Saffron

Saffron, the costliest spice is produced in Jammu & Kashmir. Production in India is estimated to be around 12 tonnes. Spain has been the major producer with about 35 tonnes, but the production has come down to about 15 tonnes now. Iran is at present the leading producer with about 80 tonnes, But Iraninan saffron is considered to be inferior in quality. India exports saffron in small quantities to a few countries.

Vanilla

Vanilla cultivation is developing in Karnataka, Kerala and Tamil Nadu. The present area is reported to be around 700 ha. Being the plantings done in the last few years present production is negligible, less than 5 tonnes. Vanilla is the second costliest spice next to saffron.

Herbal Spices

There are a number of herbs such as rosemary, thyme, marjoran, oregano, chive, sage, basil, parsley and Burmese coriander which are produced in small quantities. A part of the production is freeze dried and exported from India.

Value Added Products

Value addition has received attention in our exports especially in recent years. There are a number of value added products exported from India. While their export in 1986-87 was only 7,551 tonnes costing Rs. 25.1 crore it has increased to 20,165 tonnes costing Rs. 130.1 crore in 1991-92 and to 34,956 tonnes costing Rs. 380.8 crore in 1996-97. The annual growth is estimated at 16.59% in quantity and 31.21% in value.

The important value added products are namely, oils and oleoresins of spices, ground spices, curry powder and dehydrated and freeze dried spices. Export of oils (other than mint oil) and oleoresins has increased steadily in the last one decade. The growth is estimated at 25.04% in volume and 33.19% in value per annum. Their export was 2,004 tonnes valued at Rs. 154.5 crore during 1996-97. India commands over 60% of the supply of these products to the world market. In the recent years export of mint oil has picked up notably. Export during 1996-97 was1,805 tonnes valued at Rs. 91.2 crore. India is now the largest supplier of Japanese mint oil to the world market. Export of oils and oleoresins are mainly to USA, UK, Italy, Germany and Canada.

Curry powder is a traditional value-added export item. Export has increased from 2,712 tonnes valued at Rs. 4.1 crore in 1986-87 to 4,250 tonnes valued at

Rs. 18.2 crore in 1996-97. Besides curry powder, ground spices individually and in mixes and curry paste are exported in small quantities. Indian curry powder export is largely to Kuwait, Saudi Arabia, UK and USA.

Future Outlook

Low productivity, strong internal demand and high cost of production are the main reasons which make India uncompetitive in the international market for many spices. To increase productivity, technological interventions such as hybridisation and tissue culture are to be adopted much more intensively. A number of varieties of different spices with high yielding and good quality have been released from the research stations. Their cultivation is to be popularised with better farming techniques. On-farm processing has to be improved by transferring modern methods to farmers.

Steps have been taken for the large scale cultivation of non-traditional spices like vanilla, paprika and herbal spices. Programmes for expansion of area under cultivation of specific exportable varieties of turmeric and ginger are to be implemented regularly to increase production. Identification of spices for organic cultivation is another area where attention is to be given in a big way.

Value added products especially spice oils and oleoresins have been identified as the extreme focus area. Other products for which more attention is given are curry powder and paste, spice blends/mixes, spices in consumer packs etc. Besides, efforts are to be made to develop/commercialise new end products like micro-encapsulated spices, sterilised spices etc. for export. More and more units for the production of value added items should be set up through joint ventures with foreign collaboration for increasing exports.

Quality has become the most important parameter to trade with developed countries. In order to cope up with the situation, additional measures especially for avoiding pesticide residues and aflatoxin beyond MRL are to be taken up to improve quality of Indian spices. More training programmes are to be arranged to educate the farmers on better cultivation practices and on-farm processing. Cleaning/processing/warehousing facilities at exporters level are also to be strengthened. More and more exporters should adopt logo and spice house certificate schemes.

A laboratory with the aim of quality evaluation and upgradation of existing spices have been set up at Cochin. The activities of the laboratory is enhanced and exporters are able to take advantage of the laboratory for quality check. A regional laboratory will also have to be set up soon at Bombay which is the next important centre for spice export. The programme taken up to assist exporters to establish in-house quality labs is to be strengthened. About six units have already got accreditation to ISO 9000 now. More and more exporters should be encouraged to adopt ISO 9000 in their processing and manufacturing units.

Any export depends on three factors, namely availability, acceptability and affordability. Following these factors India must produce sufficient quantity of quality spices at competitive price for export. If so, India has bright future for spice export.

REFERENCES

Anon,1997. Area and production of spices in India and World, Spices Board, Cochin. PP 133

Anon, 1998. Spices Export Review 1997-98, Spices Board, Cochin.

Anon, 1998. Spices statistics, 4th edition, Spices Board, Cochin.

Anon,1998. Imports of spices into selected markets 1992-96. International Trade Centre, UNCTAD/WTO.

George, C. K., 1976. Pepper production in Five Year Plans, Proceedings of the International Pepper Seminar held at Cochin, Spices Export Promotion Council, Cochin.

George, C. K., 1982. Socio-economic study on production and productivity of pepper holdings in the International Pepper Community Member Countries, Economic and Social Commission for Asia and the Pacific, Bangkok.

George, C. K., 1984. Availability of pepper in India for processing. Proceedings of the Workshop on Processing of Pepper and Pepper Products conducted by the Central Food Technological Research Institute, Mysore, India.

George, C. K., 1984. World Vanilla Industry. *Indian Spices*, Vol. XXI, No.1.

George, C. K., 1989. Spices export-accent on value addition, Survey of Indian Agriculture, The Hindu, Chennai.

George, C. K., 1989. International trade in vanilla. *Spice India* Vol. II, No.2.

George, C. K., 1993. Spices exports: Present status and future outlook, Proceedings of the International Food Convention, Central Food Technological Research Institute, Mysore, India.

George, C. K.,1994. Ensuring quality in spice exports. *Forum* Vol. XXX, No.2.

George, C. K., 1994. Processing and product development of pepper-Indian experience, International Pepper Seminar, Towards A More Vibrant Pepper Economy, ESCAP, Bangkok.

George, C. K., 1994. Spice Industry—A domestic and global overview, *Indian Horticulture*, Vol. **39**, No. 3.

George, C. K., 1996. Report of the Research and Development Specialist for Ginger in Nepal, Phase I,. Secondary Crops Development Project, Ministry of Agriculture, His Majesty's Govt. of Nepal, Kathmandu.

George, C. K., 1996. Report of the Research and Development Specialist for Ginger in Nepal, Phase II,. Secondary Crops Development Project, Ministry of Agriculture, His Majesty's Govt. of Nepal, Kathmandu.

George, C. K., 1996. Report of the Supply Study and Production Guidelines for the Development of Spices Production Industry in Papua New Guinea for Export to the United States and Canada, Phase I,.Commonwealth Secretariat,London.

George, C. K., 1996. Technology Mission on Black Pepper, Mini Mission Research, Mini Mission II Production, Mini Mission III Storage, Marketing and Quality Improvement and Mini Mission IV Export Promotion, Department of Agriculture, Govt. of Kerala, Trivandrum.

George, C. K., 1997. Report of the Survey of Spices in North America and Marketing Guidelines for the Development of Spices Production Industry in Papua New Guinea for Export to the United States and Canada, Phase II,. Commonwealth Secretariat, London.

George, C. K., 1998. Strategies for Accelerating Production of Indian Spices for Export, National Horticulture Conference, 1998-99, New Delhi.

George, C. K., 1999. Research and Development of Spices in the Next Millenium in Kerala, Kerala Science Congress, 1999 Central Plantation Crops Research Institute, Kasaragod.

Nandakumar, T., 1996. The Global Spice Trade and the Unguary Round Agreements. International Trade Centre, UNCTAD/ WTO/ Geneva.

Recent Trends in Spices and Medicinal Plants Research pp. 175-183, 2000
Ed. Amit Krishna De
Associated Publishing Co., New Delhi

3.7

CULTIVATION AND PROCESSING OF MEDICINAL AND AROMATIC PLANTS—OPPORTUNITIES FOR UPLIFTMENT OF RURAL ECONOMY

D. D. Patra* and Sushil Kumar*

Abstract

Even today the rural folk as well as a sizable portion of the urban population depend on traditional system of medicine. Likewise a significant amount of allopathic drugs prescribed today contain one or more derivatives of plants. Because of varieties in soil and agroclimatic conditions, India has tremendous potential of growing medicinal and aromatic plants. India is one of the world's leading biodiversity centres with the presence of over 45,000 different plant species of which about 15,000-20,000 plants have good medicinal values. Medicinal and aromatic plant industry consists of major activities like cultivation and processing of plants, generation of number of ancillary activities like fractionation of essential oils, preparation of isolates and high value perfumery chemicals, production of attars, perfumes and cosmetics, isolation of crude drugs and establishment of herbal drug units. Depending upon the agro-climatic conditions, the medicinal and aromatic crops can give annual return ranging from Rs. 10,000-50,000 per hectare per annum. This paper highlights the economic and employment potential of these high value crops with respect to rural development.

Introduction

Plants have been a source of medicine and aroma chemical for thousand of years. Over three quarters of the world population relies mainly on plants and plant extracts for health care. Currently, there is a resurgence of global interest for "green drugs" green foods etc. all based on plant, specially medicinal and aromatic plants because of inadequate supply of drugs in certain parts of the world, prohibitive costs of certain items and side effects of several synthetic drugs in current usage.

The green rich developing countries of the Asian region have tremendous potential for exploitation of a range of their existing flora for pharmaceutical and curative purposes. They have developed and utilized them over thousand years as traditional system of medicines, Ayurveda, Unani, Siddha (India), and Jamu (Indonesia) etc. for their primary health care need. Because of its unique geographical

*Central Institute of Medicinal and Aromatic Plants (CIMAP); P. O. CIMAP, Lucknow-226 015

locations with respect to soil and agroclimatic conditions, India has tremendous potential of growing medicinal and aromatic plants. Indeed, India is one of the world's leading biodiversity centres with the presence of over 45,000 different plant species of which about 15,000-20,000 plants have good medicinal values. More than 1300 aromatic flora have also been recorded to occur in India. Besides this, several flavouring species are also produced. Out of these, 35 are commonly used, though only 17 species constitute 90% of the current world trade (Sharma, 1997).

Recent estimates indicate that the annual value of medicinal plants derived from developing countries is about $32 billion (Rs 100,000 crores) and which India's share is about 4000 crores. There are about 80 major medicinal plant species distributed to more than 40 families, which have potential medicinal values. Similarly, around 44 aromatic plants belonging to 18 families constitute the major source of essential oil in India. Over 20 species are cultivated in areas ranging from limited to 60,000 ha. Besides these, around a dozen other species have future prospects for commercial utilization.

The medicinal and aromatic plant industry consists of major activities like cultivation and processing of plants and generation of number of ancillary activities like fractionation of essential oils, preparation of isolates and high value perfumery chemicals, production of *attars*. Perfumes, cosmetics, agarbatties and establishment of herbal drug units. All these activities being highly labour intensive there is tremendous scope of employment generation and poverty alleviation in rural sector. Equally potential is the development of medicinal and aromatic plants based small scale industries to uplift rural economy.

Depending upon the farm management practices, crop cultivated and prevailing soils and agroclimatic conditions, the medicinal and aromatic crops give annual return of Rs. 10,000 to 50,000 per hectare per annum. In some aromatic crops it is even more. Many of the medicinal and aromatic plants grow in marginal lands like salt affected soil and with saline irrigation water (Misra 1987, Singh *et al.,* 1987, Patra and Singh, 1995).

Scope of Technology Inputs in Rural Development

Cultivations

As a result of R&D activities being carried out by CIMAP, a few CSIR and ICAR Institutes and some universities for the last 3-4 decades, agrotechnologies for a large number of medicinal and aromatic plants have been developed. CIMAP has been instrumental in developing suitable cultivars and standardizing the agronomic practices of several medicinal and aromatic crops suitable for specific agroclimatic conditions. More than 20 improved varieties in 10 medicinal plants and 40 varieties in 18 aromatic plants have been developed and released for cultivation in India.

Mints

Four species of *Mentha* are commercially cultivated in India. These are Japanese mint known as menthol mint *(Mentha arvensis)*, pepper mint (*M. piperita*), bergamot mint *(M. citrata)* and spearmint *(M. spicata)*. These four species are now the main sources of aroma compounds viz. menthol, carvone, linalool and linanyl acetate used widely in food cosmetics and pharmaceutical industries. Given in Table 1 are some statistics of area and production for various commercial mints in India (Anonymous, 1997).

TABLE 1: AREA UNDER CULTIVATION AND PRODUCTION LEVELS OF OILS OF VARIOUS MINT SPECIES IN INDIA IN THE YEAR 1996

Name	*Area under cultivation (ha)*	*Production India (tonnes)*	*Total world production (tonnes)*	*Major producing countries*
Menthol mint	50,000	8,000	15,000	India, China Brazil USA
Pepper mint	16,000	100	2,000	USA, France Brazil, India
Bergamot mint	1,200	90	200	USA, Brazil
Spear mint	2,000	150	1,500	USA, China

The cultivation of mints in this country dates back to about 300 years. It has progressively become popular to the farmers of the northern plains. The main reason for the continuous increase in mint cultivation lies in the internal consumption and demand for menthol and mint oils from other countries viz. USA, Germany, France, U.K. Singapore, Brazil, Australia and Japan. Another reason is that the mint crop can be taken as a bonus after the main *rabi* crops sucn as wheat, potato, toria, gram etc. have been harvested. The accommodation of mint in the traditional rotations has encouraged more and more farmers to improve their remuneration as a bonus crop. Due to this, the export of mint oil specially menthol and its derivatives has increased several fold during the last five years (Table 2). Furthermore as per the Chemical Marketing Report (1996) since China, which has been the leading producer and exporter of mint oil has reduced its production, India has lots of scope in capturing the world market in exporting mint oil and menthol and it can be a good competitor to US market.

TABLE 2: EXPORT OF MENTHOL MINT OIL AND ITS DERIVATIVES FROM INDIA

Item	1990-91		1994-95	
	Quantity (Tonnes)	*Value* (Rs.in lakhs)	*Quantity* (Tonnes)	*Value* (Rs. in lakhs)
Menthol mint oil	58	100	586	1894
DMO	19	27	527	1110
Menthol	420	956	690	2890

In India, priority should be given to produce food, to provide food security for its people. Accordingly the cultivation of mints at the expense of food production has to be minimized. The rotation of mints with food crops in *Rabi* and *Kharif* seasons helps in the control of pests and diseases of all the crops. The rotations are possible both with normally planted mints and transplanted mint. Similarly mint as a minor crop can be intercropped with some major crops and some minor crops can be intercropped with mint as a major crop. Mint sugarcane intercropping has been established to be the most profitable cropping system both in tarai and northern Indian plains (Kothari *et al.*, 1987). Growing of mints with garlic and onion as well as food crop like wheat and sunflower can be done for higher per capita profitability. Being an agro-based industry, it generates lot of employment in the rural sector. Available estimates indicate that it generates about 12 million mandays for unskilled, semi-skilled and skilled labourers in its cultivation, processing and value addition.

Mint has been observed to be a potential intercrop in sugarcane for effective use of height, moisture and nutrients. Among different species, Japanese mint, bergamot mint and spearmint were found most suitable for spring planted sugarcane. Although intercropped mint affected the tillering of sugarcane, the reduction was however, compensated by higher individual cane weight. Mint as intercrop increased the Land Equivalent Ratio (LER) from 1.62 to 1.85. Intercropped the *M. arvensis* and *M.spicata* gave oil yields of 106 and 132 L/ha giving net return of Rs. 40,000 and 50,000 respectively against Rs. 33,000 under sole cropping of sugarcane (Kothari *et al.*, 1987a).

Aromatic Grasses

Aromatic grasses belonging to *Cymbopogon* spp. and *vetiver* are potential sources of essential oils which are amply used in perfumery and cosmetics. Citronella *(Cymbopogon winterianus)* is a rich source of aroma chemicals such as citronellol, geraniol and hydroxy citronellol. Similarly, palmarosa (*Cymbopogon martinii)* is a source of high geraniol and lemongrass *(Cymbopogon flexuosus)* of citral and are used in perfumery and flavouring industries. The unique features of these crops are that these can be grown in low fertility marginal lands, including highly eroded lands and salt affected soils with minimal inputs. Crops like palmarosa, lemongrass and *vetiver* can withstand soil pH upto 10.0 and soil salinity as high as 12 dSm^{-1} and sodicity upto ESP 55 (Patra and Singh, 1995). Since, these are marginal in nature, they do not pose any threat to the principal agricultural crops but give economic return from culturable wastelands much more than traditional agricultural crops (Table 3). It has also been observed that per hectare net return from these crops are much more than common crops grown in rainfed conditions.

TABLE 3: ESTIMATED AREA UNDER CULTIVATION, NET RETURN AND EMPLOYMENT GENERATION UNDER AROMATIC GRASSES

Aromatic grasses	*Area under cultivation (,000 ha)*	*Net return Rs/ha/yr.*	*Employment generation (million man days/year)*
Lemon grass	10	15,000	1.5
Citronella	10	25,000	2.0
Palmarosa	3	20,000	0.5
Vetiver	3	25,000	0.5

Cultivation of Flowers

India has a long history of cultivation and production of flowers commercially. The major among them are Jasmine, Rose, Marigold, Chamomile, Lavender, Gladiolus, Tuberose and Pyrethrum. The CIMAP has developed agrotechnologies and processing techniques of flower crops like Pyrethrum, Rose, Chamomile and Lavender which have been found to be highly remunerative (Table 4). Farm operations involved in cultivation and processing of these crops require lot of labour specially in plucking of flowers. So, besides higher economic return, there is lot of scope in employment generations in rural sector specially for women who are specially trained for specific farm activities. Development of suitable agro-industries for processing of the flower produce and value addition could further enhance the employment generation and per capita income.

TABLE 4: COST OF CULTIVATION AND ECONOMIC RETURN FROM DIFFERENT FLOWERS

Crops	*Expenditure per hectare/year* (Rs.)	*Production per hectare*	*Total income* (Rs.)	*Net return* (Rs.)
Rose	50,000	30-50 Q (1 lit oil)	1,25,000	75,000
Chamomile	25,000	10Q dry	50,000	25,000
Pyrethrum	10,000	500 kg	40,000	30,000
Lavender	10,000	60 lit oil	60,000	50,000

There are some other lesser known aromatic crops of which agrotechnologies have recently been developed, are found to be highly remunerative.

Common name	**Bot. name**	**Uses**
Geranium	*Pelargonium graveolens*	High grade perfumes
Clarysage	*Salvia sclarea*	High grade perfumes
Davana	*Artimisia pallens*	Food and flavour
French basil	*Ocimum basilicum*	Perfumery
Patchouli	*Pogostemon patchouli*	Fixature perfumery
Hops	*Humulus lupulus*	Food flavouring
Coriander	*Coriandrum sativum*	Food flavouring
Lavender	*Lavendula officinalis*	Soap, perfumery Cosmetics
Eucalyptus	*Eucalyptus citriodora*	Source of citronellol
Linaloe	*Bursera delpechina*	Perfumery

Cultivation and processing of some medicinal crops have also been established to be highly profitable and it helps in employment generation in agricultured sector (Table 5).

TABLE 5: ESTIMATED AREA UNDER IMPORTANT MEDICINAL PLANTS AND EMPLOYMENT GENERATION

Crops	*Area under cultivation (,000 ha)*	*Employment generation (million man days)*
Senna	25	7.5
Pappy	25	6.0
Psyllium seed	27	5.0
Others	20	5.0

Medicinal Plants

Common name	**Bot. name**	**Uses**
Medicinal Yam	*Dioscroea* spp.	Steroidal drugs
Sarpgandha	*Rauwolfia serpentina*	Antihypertensive
Pyrethrum	*Chrysanthemum* spp.	Natural insecticides
Opium poppy	*Papaver somaniferum*	Morphine, Codeine Papaverine
Artemisia (Quinghao)	*Artemisia annua*	Cerebral malaria
Ergot	*Cleviceps purpurea*	Hypertension, post partum haemorrhage
Periwinkle	*Catharanthus roseus*	Anticancer, antihypertension
Egyptian henbane	*Hyoscyamus muticus*	Tropane alkaloids
Cork wood	*Duboisia myoporoides*	Tropane alkaloids
Belladona	*Atropa belladona*	Tropane alkaloids
Ammi	*Ammi majus*	Leucoderma, sun tan lotions

The Institute has developed a high yielding strain of *guggul (Commiphora weightii)* which contains higher percentage of *guggul* lipids. An integrated agrotechnology of guggul with various medicinal and aromatic plants under irrigated and rainfed conditions has shown that it is possible to get a net return of Rs. 38,000 in second year, Rs 44,000 in third year, Rs 48,000 in fourth year and about Rs. 70,000 in fifth year depending upon the intercrops.

Processing Technologies of Medicinal and Aromatic Plants

As a result of R&D efforts by CIMAP and few other institutes a number of improved distillation and processing technologies have been developed. CIMAP has developed improved distillation plants for most of the important essential oils (Singh and Thakur, 1995). The units are low cost and energy efficient field distillation plants which are easy to operate. Available estimates indicate that in mint growing regions about 30-40 percent growers have their own distillation units. CIMAP, Lucknow in collaboration with Spices Board, Cochin is going to develop an improved distillation unit for the farmers. Apart from distillation of oil, some small scale industries have been established for menthol crystallization which adds value to the main produce thereby enhance farm return and generate employment to skilled and semiskilled labourers. Crystals are of two types: bolds and flakes. Bold crystals are of high demand in the industry and it fetches 20-25% more prices than the flakes. About 60-62% of crystals can be separated from the improved varieties which as such contain about 80% menthol. After the separation of menthol, dementholized oil (DMO) can be as such sold in the market or can be processed further to produce terpens, menthone or liquid menthol.

Tremendous development has occurred in the processing of rose oil. The rose oil pilot plant based on CIMAP technology, installed in Aligarh district has proved to be boon to the local rose growers who now get better recovery of oil as compared to the country made stills.

Some of the processing technologies developed for some important medicinal and aromatic plants could play a significant role in development of agro-based industries.

Product	Raw materials
Menthol	*Mentha arvensis*
Diosgenin	*Dioscrea* sp. (rhizome)
Menthone, monoterpenes	DMO
Citronellol	Citronella oil
Geraniol	Palmarosa oil
Hyoscyamine & hysoscine	*Duboisia* sp., *Hyosyamus* spp.
Colchicine	*Gloriosa superba* (tubers)
Jasmine concrete, absolute	Jasmine flowers
Tuberose	Tuberose flowers
Pyrethrum oleoresin	Pyrethrum flowers
Total alkaloids, ajmalicine	Catharanthus roseus
Vetiver oil	Roots of vetiver
Rose oil	Flowers of *Rosa domascena*
Nagarmotha oil	*Cyperus scarious*
Sennosides	*Cassia* spp.
Citral	Lemongrass oil
Carvone	Spearmint oil
Linalool and linalyl acetate	Bergamot mint
Artemisin	*Artemisia annua*
Arteether	*Artemisinin*
Taxol	*Taxus baccata*

Impact on Rural Development

Most of the medicinal and aromatic plants give an annual profit of Rs. 15,000-50,000 per hectare. In some cases the profit is even more. So, in one hand, it increases the farm income and engages jobless people in its cultivation and processing. On the other hand, development of agro-industries opens avenues for business as well as employment for rural people who do not find job during the off season. Extension activities play a major role in creating awareness among farmers and entrepreneurs. A spectacular example is, of Japanese mint, a natural source of menthol which has now spread to more than 50,000 ha of land, with area increasing in U.P. and M.P. Such a proposition has generated ample opportunities to million of people in rural sector. In the year 1995-96 about 8,000 tonnes of oil valued at Rs 500 million was produced which was not only being used in the country but a part was also exported. Similarly, with the development of improved varieties of menthol mint, spearment and bergamot mint, production level has increased five to ten folds during the last five years providing job opportunities to growers, processers and traders.

Promotion of cultivation and processing of medicinal and aromatic plants constituted one of the major thrust areas under the societal mission. The main objectives of the mission are:

a) Alleviation of poverty in rural sector through employment of rural and tribal people in the cultivation of medicinal and aromatic plants.

b) Additional source of income through cultivation and processing of medicinal and aromatic plants for the rural and tribal population.

c) Utilization of marginal lands through non-conventional crops.

Implementation of the programme has been possible under a three pronged approach as follows:

(i) Demonstration methods for cultivation of medicinal and aromatic plants according to agroclimatic zones with respect to cropping pattern.

(ii) "Training of trainers", who would disseminate the know-how to the actual users of the agrotechnology or processing technology.

(iii) Setting up demonstration units in tribal and rural belts to act as model farms for the local population through the help of Khadi Gram Udyog Board, Ambedkar villages and Gandhi Gram Projects and other various government and non-government organisations.

Central Institute of Medicinal and Aromatic Plants having its headquarters at Lucknow and its five field stations located at Bangalore, Pantnagar, (UP), Bonera (J&K), Kodaikanal (TN) and Boduppal (A.P.) is not only developing appropriate agrotechnologies for the cultivation and processing of medicinal and aromatic plants, but also serving the cause of growers and entrepreneurs for the rapid popularization of various technologies. Various services offered by CIMAP to growers and entrepreneurs are (Anonymous, 1997 a):

(i) Survey of land for cultivation and preparation of feasibility reports.

(ii) Supply of seeds and planting materials.

(iii) Supply of farm bulletins describing package of practices.

(iv) Assistance to growers in marketing of produce.

(v) Testing and analysis of crude drugs and essential oils.

(vi) Imparting training to growers entrepreneurs and researchers.

(vii) Supply of engineering designs for pilot plants, improved field distillation units etc.

(viii) Consultancy/technical services for cultivation and processing.

(ix) Transfer of process know-how developed.

Technologies pertaining to cultivation and processing of medicinal and aromatic plants offer immense potential for socio-economic transformation of people in the rural areas. A large number of appropriate cultivation and processing technologies have been developed. It is essential that the technologies are popularized among the masses and required technical know-how and credit facilities are made available to the users. For this a multi-organisational approach is needed. Several government departments/organisations, such as State Department of Science and Technology, Agriculture, Forest and Horticulture, Wastelands Development Board, Vikas Nigams, and voluntary organisations can play an active role in the popularization of technologies and transfer of basic know-how. Availability of credit facilities to the growers and entrepreneurs on easy term by nationalized banks, cooperatives, NABARD, Small Industries Development Bank of India (SIDBI) and Khadi and Village Industries Commission (KVIC), will certainly inspire more and more people to venture into the project. NCERT, New Delhi, has recently developed courses for vocational education in medicinal plant industry. Various state governments and administration of union territories should popularize the course, so that sufficient trained manpower is available for the mission.

References

Anonymous, 1997. *Cultivation of Menthol Mint in India,* CIMAP, Lucknow, pp. 3.

Anonymous, 1997. *Strategy Document of CIMAP*, Lucknow, pp. 5.

Kothari, S. K., Singh, V., 1987. Intercropping of mint species in spring planted sugarcane under tarai conditions of Uttar Pradesh. *Indian J. Sugar Tech.,* **4** : 1-6.

Kothari, S. K., Ram, P. and Singh, K., 1987. Studies on intercropping of autumn planted sugarcane with bergamot, pepper and spearmints in tarai areas of Uttar Pradesh. *Indian J. Agron.* **32** : 406-410.

Misra, P. N., 1987. German Chamomile *In : Eco Development of Alkali Lands, Banthra-A Case Study*. T. N. Khushoo (ed.), NBRI, Lucknow, pp. 88.

Patra, D. D. and Singh, D. V., 1995. Utilization of salt affected soils and saline/sodic irrigation water for cultivation of medicinal and aromatic plants. *CROMAP* **17** : 378-38.

Sharma, J. R., 1997. Industrial utilization of medicinal and aromatic plants in Asia. ANMAP's Symposium and Regional Export Consultation—A summarized report. *JMAPS* **19** : 753-764.

Singh, A., Singh, M. and Singh, D. V., 1987. Vativer—a promising crop for problem soil. *Pafai J.,* **2** : 27-29.

Recent Trends in Spices and Medicinal Plants Research pp. 184-188, 2000
Ed. Amit Krishna De
Associated Publishing Co., New Delhi

3.8

UNANI FORMULATIONS AND THEIR QUALITY CONTROL— DEVELOPMENT OF A MODIFIED PYKNOMETER

Shamim A. Siddiqui*

Abstract

Modern or allopathic medicines generally consist of one or a few pure synthetic or natural substances whereas indigenous medicines of Ayurvedic or Unani systems mostly have a number of crude or semipurified drugs of herbal, mineral or animal origin. As such, standardization of indigenous polypharmaceuticals becomes a challenging analytical exercise. An outline of complexity of the problem as well various physical and chemical parameters which have been under studies for standardisation of Unani formulations has been described.

Introduction

Medical ethics demands that every drug formulation must be prepared by using genuine ingredients in their correct proportion and adopting the correct procedure so that the consumer could get authentic medicine for the alleviation of his suffering. Few decades ago pharmaceutical recipes were being religiously prepared either by the physicians themselves or by their trained and sincere pharmacists. Those persons were conversant in recognising the ingredients as well as in the art of preparing polypharmaceutical formulations. As such, standardization of a finished drug product was seldom felt in those days. The fast and easy going life of today led to the emergence of patent drug recipes manufactured commercially by using the tested and trusted formulae and well defined methods of renowned *Hakims* and *Vaids* described in their medical treatises, viz *Qarabadins* and *Nighentoos*. Therefore, it became imperative that in order to effectively check any advertent or inadvertent adulteration, substitution or deletion of some costly or rare ingredient or to effectively check a fault in the method of processing, each and every finished product must be properly tested and standardised for certain defined parameters before the marketing of the product.

*Research Officer, Drug Standardization Research Unit, RRIUM Campus, Kursi Road, Lucknow, U.P.

We know that allopathic medicines mostly consist of a few pure synthetic or natural substances. The standardization of these medicines for their quality control is not very difficult. It is routinely carried out through estimation of one or more important chemical compounds present in the formulations. On the contrary, indigenous medicines generally consist of a large number of crude or semipurified drugs of herbal, mineral or animal origin and in most of the cases qualitative or quantitative determination of the ingredients is next to an impossible task. As such, for controlling the quality of indigenous formulations an indirect approach consisting of study of a large number of physical and chemical characters of such drugs have been evolved (Tables 1-2). The parameters shown are being studied at CCRUM standardisation laboratories for the various unani recipes prepared with genuine ingredients in the proportion and methods prescribed in famous unani formularies. Bulk density of a formulation is one of its important physical parameters. It is the mass of a unit volume of the drug determined without disturbing its texture. A deviation in its value from the standard one gives an idea of not only the presence of genuine ingredients in their correct proportion but also of the fact that a tablet has been prepared under desired pressure and whether a syrup is of desired consistency.

TABLE 1: STANDARDISATION OF INDIGENOUS DRUG FORMULATIONS

* Separation of the ingredient not possible in most of the cases.
* Identification of the ingredients also not possible.
* Standardization procedures used for allopathic medicines can not be used for herbo-mineral formulations.
* Indirect procedures or finger printing procedures are adopted.
* The method adopted should be so simple that it could be carried out with the apparatus and equipment generally available in a chemical laboratory.

A simple apparatus, bulk density pyknometer, had been developed by our unit few years ago. In that apparatus a fine capillary was fused to the bulb of a 25 ml volumetric flask with a paper scale fixed in the background for noting rise of liquid in the capillary on addition of the drug. The same has now been simplified replacing the volumetric flask with a 10 ml good quality measuring cylinder (Fig. 1).

Materials

1. The modified pyknometer
2. Double filtered mustard oil
3. Chemical balance with weight box

Method

(1) The pyknometer is carefully filled with mustard oil without its adherence to the walls, a little above the bend of the capillary, so that the meniscus comes to a convenient mark (say r_1).

(2) The assembly is then weighed on a chemical balance say w_1 gm

(3) One or two ml more mustard oil (say v ml) is poured into the assembly and the rise of meniscus inside the capillary (say r_2) is noted.

(4) For every one ml increase in the volume the rise of the liquid meniscus inside the capillary, the calibration factor (C.F.) is calculated as follows.

$$C.F. = \frac{V}{r_2 - r_1}$$

(5) A small amount of the drug, nearly 1-2 ml in volume, is carefully introduced into the assembly. The rise of meniscus inside the capillary (say r) is noted and the assembly weighed again (say w_2 gm).

(6) The exact volume of the drug (say V ml) is calculated as follows:

$V = C.\ F. \times r$ ml

(7) The bulk density of the drug is then calculated as follows:

$$\text{Bulk Density} = \frac{\text{Mass}}{\text{Volume}} = \frac{w_2 - w_1}{v}\ g/cc$$

TABLE 2: PARAMETERS OF STANDARDISATION

	Physical		Chemical
1.	Colour	1.	Fat content
2.	Smell	2.	Volatile oil
3,	Taste	3.	Total ash
4.	Chipping	4.	Water soluble ash
5.	Average weight	5.	Acid insoluble ash
6.	Bulk density	6.	Saponification value
7.	Loss of weight on drying	7.	Acid value
8.	I. P. solubility in water	8.	Iodine value
9.	I. P. solubility in alcohol	9.	Crude fibre
10.	Successive extractive values	10.	Tannins
	(i) Petrol ether	11.	Resin
	(ii) Chloroform	12.	Sugar content
	(iii) Alcohol		(i) Reducing
11.	pH values, 1% and 10% solutions		(ii) Non-reducing
		13.	Nitrogen content
		14.	Phosphorous content
		15.	Total alkaloids
		16.	Various metallic content
		17.	Thin layer chromatography
			(i) Petrol ether extractive
			(ii) Chloroform extractive
			(iii) Alcohol extractive

Results and Discussion

Rise in capillary meniscus on adding 2 ml of mustard oil = 2.6,

$$\text{Calibration Factor C. F.} = \frac{2}{2.6} = 0.77$$

S.No.	*Drug*	*Mass (gm)*	*Cap. Rise*	*Volume (cc)*	*Bulk Density gm/cc*
1	HKN	1.8332	1.5	1.155	1.60
2.	QTQ	2.2122	2.0	1.540	1.43
3.	HM	0.9620	0.9	0.693	1.38
4.	MHY	1.6624	1.5	1.155	1.44
5.	JSQ	1.1432	1.1	0.847	1.35
6.	STS	1.4806	1.4	1.078	1.37

The presently developed pyknometer utilises only 10 ml of the liquid as against 30 ml utilised by the former one. It is also easy to clean and therefore, several determinations can be done in a day. The values of bulk density determined by this apparatus are accurate to the second place of decimal.

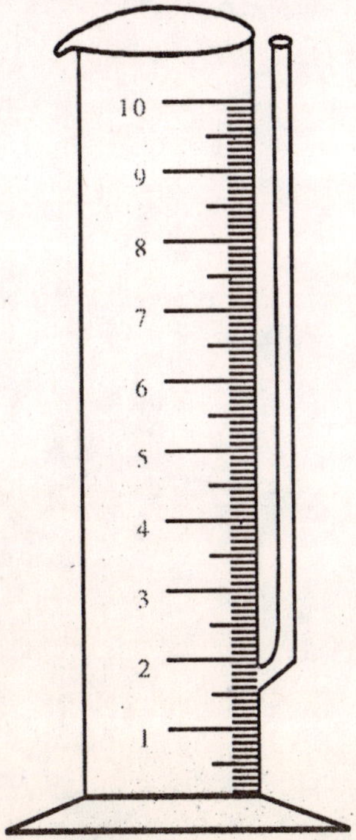

Fig. 1. Modified Pyknometer

Precautions

1. While noting the reading of meniscus the assembly must be placed on a perfectly horizontal base.
2. While introducing the oil or the drug into the apparatus, care should be taken to avoid adherence of the material to the stem of the assembly.
3. Care must be taken to avoid overflowing of oil from the assembly.
4. If some fat (oil or ghee) has been used in the preparation of the drug then mustard oil should be replaced by some hydrophillic liquid of lower density. For example Butanol (Density 0.8 gm/ml.)